ECOLOGY AND EARTH HISTORY

Ecology and Earth History

RICHARD N. T-W-FIENNES

CROOM HELM LONDON

First published 1976
© 1976 by Richard N. T-W-Fiennes
Softcover reprint of the hardcover 1st edition 1976

Croom Helm Ltd.
2-10 St. John's Road, London SW11

ISBN-13: 978-94-011-6378-1   e-ISBN-13: 978-94-011-6376-7

DOI: 10.1007/978-94-011-6376-7

CONTENTS

INTRODUCTION

This volume outlines in simple terms the basic concepts of ecology. It is not a textbook of ecology, but an introduction to the subject and to the more advanced presentations of special aspects that will appear in succeeding volumes. There are many excellent textbooks of ecology, to some of which the reader is referred and to which the present volume should also serve as a useful primer.

I have linked this presentation of ecological theory to a survey of present theories of Earth History. All of earth history is, in a sense, ecology. Earth's environments have evolved throughout the ages from the time when no life existed, just as life forms have evolved from primitive to complex. It seems to me, therefore, that one should understand how these environments have come into being. I have also presented a simple account of the energy systems that are used by living things. These too need to be understood, if we are to appreciate energy problems. Man, like all other living things, requires the energy provided by his food, but unlike other creatures his way of life demands additional ('auxiliary') energy, and this has introduced a new factor to man's ecology. Hitherto, most of this auxiliary energy has been obtained from fossil fuels, coal, oil and uranium. But these are not inexhaustible; they represent capital, not income, and it is still not clear what the sources of the auxiliary energy will be.

The sun's energy is responsible also for the movements of the clouds, the winds and the ocean currents, and for circulating fresh water from the oceans to land surfaces. Otherwise, there could be no life on land. Here again, man's ecology demands more than that of other animals. Not only must man have water for drinking, but also for removal of his industrial and urban wastes. There are signs too that, lacking further energy input, this resource may be inadequate.

Man's future is beset with many doubts and complexities. To overcome them, mankind must understand his nature and move forward to find solutions that will be of benefit to all.

NOTES

1   J.C. Wylie, *Fertility from Town Wastes,* Faber, London, 1955
2   H. Godwin and A.G. Tansley, *The Natural History of Wicken Fen,* Bowes & Bowes, Cambridge, 1929
3   R.N.T -W - Fiennes, *The Ecology of the Grasses of the Lango District of Uganda, East Africa,* Government Printer, Entebbe, 1939

4 H.W. Mulligan (ed.), *The African Trypanosomiases,* Allen & Unwin, London, 1970
5 R.N.T -W - Fiennes, *Man, Nature and Disease,* Weidenfeld & Nicolson, London, 1964

Further Reading

Bertram, Colin, *Adam's Brood,* Peter Davies, London, 1959
Cloudsley-Thompson, J., *Animal Conflict and Adaptations,* Foulis, London, 1965
Reed, Laurence, *An Ocean of Waste,* Conservative Political Centre, London, 1972

1 GENERAL PRINCIPLES OF ECOLOGY

The Evolution of Habitat

In its simplest meaning, ecology is the study of habitat (Greek *oikos,* house or habitation). Ecology has, however, become a complex science with a terminology of its own and journals devoted to it. Only recently has the word appeared in the popular press in connection with environmental problems. In spite of this, as a science it has a respectable ancestry. Early naturalists, including Wallace and Darwin, recognised the relationships of living things with their habitat and with each other. Definitive studies on ecology appear to have been initiated by a German scientist, Möbius, in the 1880s. Subsequently and until the present day, the importance of such studies has been widely recognised and they have been pursued in Britain and North America, as well as in Germany. A critical review of the subject is given by Macfadyen,[1] and a short, but useful account by Healey.[2]

It is obvious that certain kinds of animals or plants can only live in certain types of habitat. Fishes cannot live out of water and a monkey thrown into water will drown if it cannot escape. It is generally accepted, too, that since life appeared on earth there has been a progression of life forms of increasing complexity. This progression is attributed to evolution, a process which increasingly adapts living things to be successful in the areas where they are evolved.

It may be less obvious that not only life forms undergo a process of evolution, but habitats as well. You have only to observe the local duck-pond to see that, left unhampered, water weeds grow in it and in time it gets filled and becomes earth with grass and trees and shrubs growing where there was once water. Similarly, look at any disused road or air strip and you will see that the surface is gradually weathered away, plants appear, in time soil is formed and it will eventually appear not unlike the surrounding area. Even on hard concrete, you can see that lichens quickly appear on the surface, followed by mosses and other plants.

Ecology basically looks at habitats in an evolutionary sense. It regards them initially as sterile areas, devoid of vegetation or other forms of life. 'Pioneer' forms of plant life then appear and begin to erode the surface or alter it in some other way which makes it more suitable for colonisation. The process continues and more demanding plants appear, which continue the process. At the same time, weathering assists the development: wind and rainfall break up and fissure the surface; frozen water in winter widens the fissures; roots of shrubs or small trees will enter the fissures and break them wide open; and soil begins to be formed. The evolution of a habitat proceeds by a series of steps or 'successions' until the 'climax' is reached. The 'climax' is the point at which both habitat and life forms on it are in full unison, so that development will not proceed further provided there is no change of climate. Examples of 'climax' habitats are primary forest, such as the big equatorial forests in Central Africa and South America, and prairie grasslands like those developed on chernozem soils across central North America and the steppes of Siberia and southern Russia into Europe. In such areas, a characteristic animal fauna is developed also, adapted to the conditions of the habitat: there are forest animals and birds in the forests, and plains animals and birds on the plains, differing in their ways of acquiring food, in colour and other protective measures, and in their general reactions and physiology.

The general process of 'succession' is known as a 'sere'. The process arising in water is known as a 'hydrosere' and that on land or rock surfaces as a 'xerosere'. The direct succession to the climax is known as the 'primary succession'. If, however, there is interference with the climax, as by cutting timber or clearing for agriculture, a 'secondary succession' develops, which may be a reversion to an earlier stage of the 'primary succession', or something different. When there is interference with a pre-climax stage of a succession, as, say, by dry season burning, the 'primary succession' may be deflected, and we speak of a 'deflected succession', which may reach a kind of climax of its own persisting until the cause of the deflection is removed.

An understanding of these seres and their stages is fundamental to an understanding of environmental problems and pollution by human activities. Most of these create a new form of pioneer surface, on which the object of reclamation is to recreate an advantageous succession leading to recovery in the shortest possible time. Soil erosion leads to loss of topsoil and in tropical areas to laterisation — that is chemical change and rock-like concretion — of the subsoil. The reverse process under natural conditions may take centuries, but the ecologist may find ways to hasten the process. Large sections of the Sahara carried human and animal populations within human memory and have been converted to desert largely by human activities.[3,4] If the process of desiccation can be reversed, this can only be done with a knowledge of the ways in which suitable ecological successions can be encouraged. A polluted river or estuary is another instance of a pioneer type of substrate, which can be remedied if the right kind of microbes are introduced to devour the pollutants or in the case of chemical pollutants to alter them by their metabolic activities. Even oil slicks can be

attacked by the right kind of organisms and quickly degraded
into harmless substances or even into nutrients for higher forms of life.

The spate of journalism in recent years on ecology and the
environment appears to ignore the absolute necessity of studying
the ways in which natural processes may be stimulated and accelerated.
Governments appear mesmerised by the necessity of appearing to
do something, even the wrong thing, rather than to invest the public's
money in fundamental research on these problems. One might suppose,
from what one reads, that human activities are universally detrimental
to earth's environment, certainly from his own point of view. The
natural successions have been either deliberately or unwittingly
'deflected'. The problem is to understand the nature of these
deflections and to remove their undesirable features; this means study.

The Ecological Community

Hitherto, we have considered the successions which occur in an ecological
sere. Let us now look at the concept of the 'community' or 'ecosystem',
to use the term introduced by A. G. Tansley[5] in 1935. First, what are
the factors which determine the type of community to be developed in a
given habitat? There are only two of major importance, the climate and
the soil. It might be objected that this ignores the importance of the
atmosphere. The composition of the earth's atmosphere has, however,
been so constant over millions of years that we can for the present
ignore it.

Like much in ecology, the influence of climate is superficially obvious.
In the most extreme climates, waterless desert and polar ice, life cannot
exist, although abundant unicellular protozoa live in the waters below
the polar ice. At the other extreme, life is most abundant and prolific.
in the wet tropics, where rainfall is heaviest and the sun's rays are most
powerful. In such areas, too, seasonal variations are at a minimum,
whereas in other areas rainfall and temperatures are variable throughout
the year and the vegetation varies also. Both plants and animals show
adaptations to lean periods when water is scarce or the temperatures low.
Trees shed their leaves; annual plants bury their seeds in the ground to
grow again when conditions are favourable; some birds — mammals too —
migrate, and some mammals, birds and reptiles hibernate or aestivate.
These are commonplace facts and do not require elaboration. They are,
however, facts of which ecology takes account, and which the ecologist
must study if he is to understand his area.

The facts relating to soil conditions are somewhat more complex.
As a basic principle, soil is formed from a parent rock, usually the rock
underlying it. However, it may be impoverished by leaching or erosion,
or enriched by wash or silt. Usually the chemicals present in a soil will
only be those present in the parent rock. If this is deficient in any of the

minerals important for plant growth, these will be deficient in the soil too. There are many minerals important for plant growth, such as calcium, phosphorus, magnesium, zinc, cobalt, molybdenum, manganese, iron and copper. In some cases, only trace quantities of the element are needed. In some areas, plant growth is limited by these deficiencies and is less luxuriant than climatic conditions would appear to warrant. As a result of the poor growth of herbage, the development of a rich soil is either retarded or inhibited. It also happens that soil minerals may be adequate for the development of a rich plant cover, but are inadequate for the nutritional requirements of animals that feed on them. In this case, the fauna is scanty, although the vegetation is rich and might be expected to support large numbers of animals. In all parts of the world, areas are known where domestic animals do not thrive on the pastures, suffering from deficiency diseases known by various names, such as pine, swayback, staggers, and so on. These diseases were for a long time a mystery to stock-breeders, until it was discovered that there existed underlying defects in the minerals present in the soil. The deficiencies are sometimes slight, and one Australian farmer remedied a deficiency of copper by drawing a short length of copper pipe across his pastures behind a tractor.

As soils develop, they become enriched by the addition of humus. This is derived from broken-down organic debris of dead plant and animal material. Thus, the more quickly rich vegetation becomes established in a habitat, the more quickly will the soil become improved and move towards climax conditions. Here again the ecologist, and indeed the town-house gardener, can accelerate the process by thorough cultivation and by addition of materials to the soil to remedy deficiencies and improve its structure and water-holding capacity. These activities, however, must be properly orientated and based on proper soil analysis and examination. With the more difficult types of habitat, such as sand dunes and salt marshes, much can be done also by introducing pioneer plants adapted to them; certain types of grass will grow on sand dunes and, when introduced, will initiate the soil formation process.

Quite apart from the plants we can see, the soil itself teems with life. Everybody knows of the beneficial effect of earthworms in aerating and improving the soil structure. The soil is also alive with bacteria, fungi, small helminths and other organisms, whose activities are advantageous; indeed, without this active life in the soil itself, many of the higher plants we can see could not exist. Not least amongst these are the small fungi and bacteria which form root nodules on coniferous trees and leguminous plants, such as clover, so fixing atmospheric nitrogen, and their allies in the soil.

Also amongst these organisms are the so-called saprophytes,

responsible for breaking down dead organic material and making its products ready for re-use. Without this 'recycling' of organic materials, life will in time grind to a halt. Unfortunately, in the euphoria which followed the discovery of agricultural chemicals, their effect on soil communities was largely overlooked as with their effect on the insect life required for plant pollination. Today, in our large combine-harvested fields scant attention is still paid to the effects of these chemicals on the soil-living communities; pests are eradicated by chemicals and deficiencies made good by the use of artificial fertilisers. Aeration and structure of the soil are ignored and experts are asking how long such conditions can persist; belatedly land-owners are stipulating specified rest periods for their fields during which recovery can take place. But can it? Again, organised ecological studies of soil life are essential, if permanent deterioration of soils is not to take place as has happened over such vast areas of the American Middle West and elsewhere.

Soils, therefore, to achieve optimum development must not only contain all the minerals necessary for the plants and animals they are to support, but must also be properly aerated and maintained by the small organisms that live in them. These organisms form the base of the ecological pyramid, their biomass (weight of living matter) far exceeding that of the visible plants and animals which depend on them at higher levels. Many of them use energy systems quite different from higher plants and animals; some do not require oxygen; some use as energy sources chemicals that are not available to higher plants and animals. Thus they perform activities that cannot be undertaken by other life forms and in situations not available to them; for instance they are found both in oil wells and in the upper layers of the atmosphere. They provide us with our antibiotics, our wine, bread, beer and soft cheeses, and are used in many industrial processes. To understand their activities is as important as not destroying them by artificial processes in agriculture.

An ecological community, then, is an association ('associes') of micro-organisms and higher plants and animals, which live together, are dependent on each other, and contribute to the improvement of the environment to a point dependent on conditions of climate and soil. We can now study the organisation of the community and the factors which lead to its stability. This is of special importance because man's activities in one way or another have so radically altered so many habitats and communities in ignorance of their structure and indifference to the effects. This is no new problem, because man has been doing it for 10,000 years or more; it is a pressing problem because of the increase in human numbers which has made the effects more widespread.

The Stability of the Ecosystem

Most plants and animals, of which the rat, man and the housefly are notable exceptions, are very 'habitat specific'; that is to say that they are confined to habitats which suit them within rather narrow limits. Quite slight alterations of a habitat result in a different species of bird, for example. When the habitats merge or intergrade, there may be a 'tension zone' where the two species are found but are in competition. Such gradations are seen in the case of a 'hydrosere', where at one end there is open water gradually rising through marsh from sloping land to upland. In such a situation, the vegetation – and the animals dependent on it – is 'zoned' from aquatic plants through marsh plants, to lowland wet-loving plants – grass and sedge – to dry-loving upland plants and trees.

A habitat comprises numerous 'niches', and there is a life form ideally suited to fill each niche. If a niche is empty, some form will quickly appear to fill it. Niches are extremely varied. We have already seen how many varied organisms live in and on the soil. Small insects and other creatures live in cracks in the bark of trees, under stones, and in every conceivable place where they can achieve their needs for food and breeding and reasonable security from predators. The term (niche) covers not only the places in which these creatures live, but also food requirements. For example, a niche is offered by animals living in the bark of trees for predators such as woodpeckers, or monkeys with long thin fingers that can get at them. Thus in each habitat, there exist food chains or food webs, which can be exceedingly complex in rich habitats such as tropical forest or rather simple in poor habitats such as desert. Basically, the energy which is the driving force of life is trapped by plant chlorophyll and used to convert carbon dioxide (CO_2) into higher carbohydrates such as starch, sugars, and fats. The stored energy is passed from mouth to mouth, through herbivores, carnivores, predators and scavengers, until it is dissipated by processes of decay and the residual organic materials are returned to the pool to be recycled.

In fertile areas, the communities developed are complex, well balanced and stable. Obviously they cannot be stable unless the numbers of plants and animals living in the community are kept within the limits that the habitat can support. To achieve this, the natural balances are maintained at optimum levels by a number of devices which serve to limit numbers when they tend to become excessive, and to build up numbers when conditions become favourable. Either way, the result is achieved in a remarkably short time. The underlying principle of this control is simple. More young are produced than can survive and the excess are eliminated in one way or another. Since the strongest – or at least the most fitted to the conditions – survive, the tendency is for constant improvement of the stock or a change of characteristics if this is advantageous because of changed circumstances. Once this principle is understood, it is easy to see the basic difficulties underlying human population problems at the present time. It is our endeavour to have as many of our babies survive as possible, irrespective of population pressures, and we thus preserve the weak and ill-adapted as well as the strong and well adapted. In this way, we could on the face of it be contributing to deterioration of the race. Authorities, such as the British geneticist Dr A.R. Penrose,[6] think that this is not necessarily the case, and that evolution in human communities is powerfully at work, but in a different way, as discussed in Chapter 4. J.B.S. Haldane also held this view.

Elimination of excessive young, which we shall examine below, is not the only way in which numbers are controlled. Both mammalian and avian populations are very sensitive to population pressures; when populations tend to become excessive, the community enters a 'stressed' state, especially if food is short. The condition of 'stress' is undergone by all members of the community and depends on alterations in the activity of glands of the endocrine system, especially the cortex of the adrenal gland. Through complicated feedback mechanisms, the animals become psychologically affected and more aggressive, the breeding urge becomes diminished and the females when they become pregnant are inclined to abort, and in any case bear fewer young. The young, when born, are inclined to be attacked and killed, and older and sick animals are also killed. In addition, the population becomes more susceptible to endemic diseases and parasites so that there is a higher natural death rate. In some animals, a migratory instinct develops and large numbers may leave the habitat to seek other suitable areas, or perish in the attempt. The suicidal migrations of the lemmings in Scandinavian countries are the best known example.[7]

The extent to which animals produce excessive young varies greatly. It is probably greatest amongst parasites, such as ticks and helminth worms, which must find a new host by rather hit and miss methods. Thus a female tick produces millions of eggs, dying herself in the process; these all hatch into larvae, of which probably only a few score find hosts; incapable of a free-living life, the rest perish. Small, vulnerable, short-lived animals such as rats and mice tend to produce large numbers of young, of which only a few survive unless conditions are especially suitable, when a plague of them may occur. The smaller birds too produce many more eggs than are likely to be needed for the next generation of breeding stock.

A number of these excess young may die at the hands of predators or from failure to find sufficient food. There is, however, a further controlling mechanism of the greatest importance, not only to the control of populations, but also to the structure and behaviour of a population within the community. This is the 'territorial' instinct. The instinct for territory is extremely ancient and universal, at any rate throughout the vertebrate kingdom. It is present among fish as well as mammals. It is surely not defunct in man, though not universally operative. The underlying principle, as with most principles of ecology, is again a simple one. Animals do not pair and breed until they have acquired a territory, and the territory they acquire must be adequate to provide them with food, a place to rear their young, and sufficient cover for their protection. A territory once acquired is respected by neighbouring animals of the same species, and trespassers, when attacked, do not defend themselves but retire. Territories are marked by various means. Birds proclaim their dominance over a territory in their song. Mammals mostly mark their territories by means of scent produced from scent glands or in the urine as with wolves and dogs. Animals with a social organisation, as say with monkeys and apes, carry the process a stage further in the 'social hierarchy' system. This is the system of the peck order; in which both males and females arrange themselves in a social order of dominants and subordinates. Amongst horses, the subordinate males are driven from the herd or killed. Amongst baboons and chimpanzees, they are tolerated but the dominant male is usually the one to cover the females. In any case, if numbers become too great, the subordinates are driven away and become easy prey for predators.

In the jungle code, indiscriminate killing of prey animals by predators does not *normally* occur, though overkills are sometimes reported. Predators take what they require and leave the main herd unmolested. The animals mostly taken are the young, which are produced in excessive numbers in any case, and the old and the sick which are more easily caught. The removal of the latter will lead to improvement of the herd generally and is not undesirable. Human beings too, such as the Red Indians and the Eskimos, obeyed these jungle laws, and although they lived by hunting the bison and caribou, never made harmful inroads on the stock, possibly because before the advent of the rifle they lacked the means to do so. They lived in harmony with nature for many generations before the appearance of Western man, the destroyer. The wanton killing of wildlife by Western man, a destruction which in some areas still continues, is an indelible blot on the human record. This is often done because of pure wantonness; at other times it is because of fancied harm done by the animals attacked, such as the usually amiable wolves or the seals which are supposed to deplete fish stocks. This is another instance in which ecological studies are required to lay old bogies.

Ecology, then, teaches us how to live together in well-balanced communities, with benefit to the habitat. Ignorance or disregard of its teachings leads to habitat deterioration or destruction. A general principle of ecology is that the species in a community are 'habitat specific' and in general confined to their own habitat, in which they fit into the community and do no harm. Of course, some species are more habitat specific than others. Those that are too habitat specific, like the Koala Bear, are doomed to extinction when the habitat is altered by natural causes, such as climatic variation. In the long run, the most successful species are those that are not habitat specific and can adapt themselves to different conditions and to different types of food and methods of feeding. Man is such an animal and has carried his nonconformity to new lengths. He not only adapts himself to a wide variety of habitats, but in various ways adapts the habitats to his needs. To interfere with habitats, naturally developed over thousands, even millions of years, is a dangerous practice. Instances abound in which formerly fertile areas have been destroyed in Southern Europe, Africa, India, Pakistan, and America. Now inland waters, estuaries, and even the oceans are threatened. It is, of course, absurd to suppose that the clock can be turned back and that man will again submit to natural habitats. However, with proper study human activities can be correlated with the requirements of the habitat; indeed many habitats could be improved and become more productive.

At the beginning of the chapter, we studied the succession of an ecological sere, by which pioneer surfaces were converted by life successions to a fertile climax. In a sense, earth history can be regarded as such a sere developed from bare rock surfaces and water to the green and pleasant land we know. It will be relevant to our argument and instructive to study earth history in this sense, and this we attempt to do in the next chapter.

NOTES

1 A. Macfadyen, *Animal Ecology: Aims and Methods,* Pitman, London, 1963
2 I.N. Healey, 'The Habitat, the Community and the Niche', in R.N.T -W - Fiennes, *Biology of Nutrition,* Pergamon Press, Oxford, 1972
3 J.L. Cloudsley-Thompson, 'Recent Expansion of the Sahara', *Intern. J. Environmental Studies,* 2, pp.35-9 (1971)
4 J.L. Cloudsley-Thompson and Anne Cloudsley-Thompson, 'Prospects for Arid Lands, *New Scientist,* Nov. 1970, pp.286-9
5 A.G. Tansley, 'The Use and Misuse of Vegetational Terms and Concepts', *Ecology,* 16, pp.284-307 (1935)
6 A.L. Penrose, *Biology of Mental Defect,* Sidgwick & Jackson, London, 3rd ed., 1963
7 C.S. Elton, *Voles, Mice and Lemmings,* Oxford University Press, London, 1942

Further Reading

Allen, W.C., Emerson, A.E., Park, O., Park, T., and Schmidt, K.P., *Principles of Animal Ecology*, Saunders, Philadelphia and London, 1949

Elton, C.S., *The Ecology of Invasions by Animals and Plants*, Methuen, London, 1958

Elton, C.S., *The Pattern of Animal Communities*, Methuen, London and New York, 1966

Errington, P.L., 'Factors limiting Vertebrate Populations', *Science*, pp.124-304, 1956

Fiennes, R.N.T -W -, *Man, Nature and Disease*, Weidenfeld & Nicolson, London, 1964

Roberts, D.F. and Harrison, G.A. (eds.), *Natural Selection in Human Populations*, Pergamon Press, Oxford, 1959

2 THE FORMATION OF THE EARTH AND THE ORIGINS OF LIFE

The Origins of Life

The probiotic earth — that is the earth before life existed on it — differed from the earth of today in two important features. First, although there was a gaseous atmosphere, this was devoid of free oxygen. Secondly, and linked with the first, the ozone layer of the upper atmosphere was absent. Since the ozone layer traps most of the sun's ultra-violet radiation, the earth's surface was exposed to a degree of radiation which would be lethal to any form of life that exists today unless protected, as beneath water.

It is generally accepted that the earth started as a cold body formed by aggregation and concretion of spatial debris, which originated in the sun.[1] Internal heat was generated by pressure and other forces, by which deep rock formations became molten, leading to intense volcanic activity. Exhalations from this activity led to the formation of an atmosphere in which hydrogen, nitrogen and carbon gases predominated; water vapour was present but no free oxygen. Their low concentration in the atmosphere today compared with their cosmic abundance argues that the lighter gases, such as neon and helium, escaped into space from the earth's gravitational forces.

The earth's surface was under constant bombardment from meteorites, which could not burn out in an atmosphere devoid of oxygen. Their impact created pitting and cratering, adding to the effect of volcanic activity. The surface was, therefore, by no means uniform in structure, nor was it level. In depressions and craters, often very large, water accumulated and eventually seas formed. Plainly, too, climatic differences existed between the poles and the equator, contributing to the lack of uniformity, and there were great temperature differences between night and day.

Table 1 *The Origins of Life*[2]

	(Years ago)
Origin of the earth	4,500 – 5,000 million
First traces of life (bacteria)	3,300 million
Simple organisms containing chlorophyll	2,700 million
Metazoa (multi-cellular life)	600 million
Life on land	400 – 500 million
Dominance of mammals and end of age of reptiles	100 million
Appearance of man	200,000
Appearance of modern man	20,000

The earth's probiotic era lasted 1,200 million years. There was no oxygen in the atmosphere, no life, fierce ultra-violet radiation, bombardment with meteorites, and abundant volcanic activity.[3] There were rocks and seas, and the elements we know today were present in a disordered state. Yet, in the sun's radiation, there was a far more powerful energy source than is available today because of admission of the ultra-violet radiation. It is known from experiments in laboratories that such radiation can cause the combination of organic chemicals, of which the most important are oxygen, hydrogen, carbon and nitrogen, to form the so-called 'biochemicals' which are the basis of life, namely amino acids, polypeptides, proteoses and proteins. From this it is postulated that, in suitable areas and under suitable conditions, there must have accumulated masses of these biochemicals in a so-called 'primordial soup'. In the depth of this primordial soup and under conditions of protection from lethal ultra-violet rays, the random meeting of diverse biochemicals could result in the evolution of some form of primitive self-replicating chemical system and so to a living thing. The power source of an early life form could not be the sun's rays, so that it would live saprophytically on the 'primordial soup' from which it was developed. There are no real improbabilities or perplexities in this concept. Possibly, the only real difference between a self-replicating chemical system and a living system is that the latter is enclosed in a cell wall or pellicle. Such could develop initially by a condensation or drying of the outer portion of the soup around the chemical system.[4]

It is reasonably well established that such could have happened under the conditions prevailing. There is no proof that it did happen. Life could conceivably have reached earth from outer space in meteorites or even in space ships or flying saucers and found a suitable medium to propagate in the primordial soup. Two important points are evident. First, the earliest life forms could not have used energy systems requiring oxygen. Secondly, they must have lived saprophytically on pre-existing chemicals — a strong argument in favour of primordial soup. In addition, they must have been protected from the ultra-violet rays of sunlight, and it is argued as most probable that this would have occurred some five metres below the surface of water, where such protection would exist. This would explain the existence of the earliest forms of recognisable life in the seas.

Creation of the Biochemicals

Although oxygen was absent from the atmosphere, it was abundant on the earth, chemically bound in the rocks and combined with hydrogen in the waters of the seas. This oxygen was gradually released, chiefly from the waters, by photo-dissociation. However its accumulation was limited by a natural process, known as the Urey effect, to approximately 0.1 per cent of today's atmospheric pressure, because the oxygen produced by photo-dissociation shaded the underlying water vapour from ultra-violet radiation. In addition to molecular oxygen, chemically unsatisfied forms of oxygen (0 and 03-ozone) were produced near the earth's surface and the ozone formed a thin layer above the molecular oxygen. Oxygen itself as well as ultra-violet light was toxic to the protoplasm of early life forms, which were not metabolically adapted to use it. The first oxygen to appear on earth was thus a pollutant. When plant life appeared, this pollution was intensified, because the oxygen excreted into the atmosphere was increased by photosynthesis. The development of photosynthesis was one of the great turning points in earth history; to it the earth owes the oxygen of its atmosphere, the raising of the ozone layer, and the development of higher life forms dependent on oxygen for their metabolic processes.[5]

The earliest earth rocks available for study are the Pre-Cambrian of some 600 million years ago, at which time metazoan life was already present. Estimates of the age of the oldest moon rocks so far acquired give a figure of 4.6 million years, and this accords well with the estimate given of earth's age of 4,500 to 5,000 million years. There are, however, no rocks of comparable age on earth which could give a clue to its composition at that time. This is not surprising since alterations due to weathering by the elements have continuously occurred on earth, but are absent from the moon. On the other hand, the moon's gravitational field is inadequate to retain more than a fraction of the gases originally present, so that the moon's atmosphere can give no clue to the atmosphere likely to have been present on earth at that time. Whether of similar origins or not, it is evident that very different influences have affected the earth's surface from those which affected the moon's. Whereas the moon's surface has altered little in the space of 3,000 million years, the earth's has been completely metamorphosed. However, on earth the release of gases from primordial rocks led to certain elements of stability which have persisted to this day, such as the concentration of salt (chloride ions) in the sea, the pH of sea water and the carbon dioxide content of the atmosphere.[6]

The situation in relation to carbon is one that bears more detailed examination. The quantities of industrial carbon dioxide reaching the atmosphere today are far in excess of what must have reached it in probiotic times from volcanic exhalations and other sources. The additional quantities are in excess of what could be removed by plant photosynthesis. Yet, any rise in atmospheric carbon dioxide is marginal only, because of the absorptive capacity of the oceans. So great is this, indeed, that it is not even reflected in a measurable alteration in the pH of sea water.

The question of the atmospheric gases has been discussed by N. W. Pirie.[7] Pirie points out that the atmosphere now contains 12×10^{20} g of oxygen, which is 235 g/cm^2 of the earth's surface. Of fossil oxygen, consumed in making sulphate and ferric iron, there may be from 1200–1500 g/cm^2. However, significantly more oxygen must have passed through the earth's atmosphere than is now present in it, so that the total of present and past oxygen may be in the region of 1800 g/cm^2. If most of the oxygen that has passed through the atmosphere is derived from photosynthesis, there should be an equivalence between oxygen and carbon now present in reduced forms, such as coal and oil. Such an equivalence cannot be shown, unless account is taken of kerogen deposits, that is the reduced carbonaceous material in sedimentary deposits. Though the figure reached in this way at 2,000 g/cm^2 is rather too high, when adjustment is made for carbon released from methane, carbon monoxide and so on, the equivalence is very close. Another estimate reveals that the present-day plant cover could make the amount of oxygen now present in the atmosphere in two to three thousand years. Thus once chlorophyllous plants had colonised the earth, the stable atmospheric system must have been developed with great rapidity; only the saprophytes lagged, resulting in locking away a great deal of carbon but without much effect on atmospheric stability.

The third important gas in the atmosphere, with oxygen and carbon dioxide, is nitrogen. It is generally held that nitrogen was present in the primitive atmosphere in the form of ammonia; this would be analogous to the ammonia-rich atmosphere of the giant planets. The source would be volcanic gases, there being a certain amount of ammonia present in plutonic rocks. However, since nitrogen is of approximately the same molecular weight as the rare gases neon and argon, it is virtually certain to have been lost into space with them. The nitrogen in the present-day atmosphere is, therefore, most likely due to biological activity. We shall see in a later chapter how the atmospheric levels of nitrogen, oxygen and carbon dioxide are kept stable by precisely regulated cycles.

We must now take a further look at the primordial soup and the ways in which living matter might have developed in it. This will involve also a consideration of the nebulous dividing line between life and non-life. As we have seen, the earth was already over 1,000 million years old when the first traces of life became apparent. The first recognisable living organisms were bacteria. Thus, even if more primitive life forms existed before the bacteria, in all this time life had not advanced far along the evolutionary scale. During this time, provided that the basic organic chemicals were assembled together under suitable conditions, the more complex 'biochemicals', consisting of the larger molecules and polymers of carbon, nitrogen, hydrogen and oxygen, could have been synthesised by the power source of solar radiation. Accumulating over this vast period, almost every conceivable variation of chemical reaction could have occurred. Amongst the chemicals likely to have appeared would be self-replicating structures and deoxyribonucleic acid (DNA), two strands of which could by a random fit have formed a double helix with appropriate linkages. The same random process would in the course of time bring together the other essentials of life, such as the organic catalysts, the energy transformers (adenosine diphosphate (ADP) and adenosine triphosphate (ATP), and the higher polysaccharides. Brought together randomly, an organised system having the characteristics of life could have appeared, and, given the conditions believed to have been prevailing during this immense span of time, surely must have appeared.[8]

Most students of the subject believe that this is what did occur. If so, the most primitive life forms obtained their energy sources from biochemicals already existing in the primordial soup, using fermentation processes such as those of yeasts which break down glucose molecules into alcohols. The use of such energy sources requires that an otherwise stable molecule, such as that of glucose, be degraded under catalytic action; the primitive organisms, therefore, needed to produce the necessary system of catalysts, a DNA system to reproduce themselves and their catalysts, and an ATP system to direct energy to constructive channels. One further property is required of a living system, namely that the stable, self-replicating, package once developed be enclosed in a membrane, without which dispersion would occur. Cell membranes are in themselves complex structures with active metabolic properties. First, they must be physically protective of the package inside. Secondly they must admit water and nutrients selectively, while excluding unwanted substances. Thirdly, they must be able to excrete waste products. Bacteria themselves possess an outer cell wall which is inert, and an inner cell membrane which is metabolically active in these ways; that is to say that the cell membrane draws on the power sources of the cell to admit and discard materials in an organised way.

Evolution of the Cell
The evolution of the functional cell membrane must have occurred over a long period of time during which the living entity survived in a medium whose composition was similar to that within the cell itself. Energy sources in those early times were not confined to the breakdown of only one type of chemical. One supposes myriads of alternatives have been available. Amongst primitive organisms which survive to the present, very many different chemical reactions are used as power sources. This has been mentioned in Chapter 1 and will be further discussed in a later chapter. All the evidence suggests that very many metabolic systems were tried and discarded, and that those found most

Fig 1 THE BACTERIAL CELL. Section of a bacterium as seen in electron micrographs. W cell wall, P plasma membrane, R ribosomes embedded in a protein matrix, D threads of DNA. Note absence of nuclear membrane.

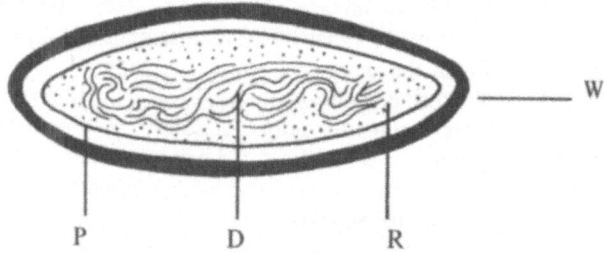

Fig 2 THE EUKARYOTE CELL. The primary divisions of the cell: nucleus, cytoplasm and plasma membrane.

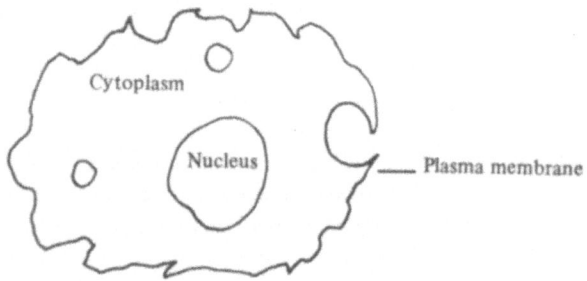

Fig 3 THE EUKARYOTE CELL: FINE STRUCTURE. Cytoplasmic structures of a cell of generalised structure as seen in electron micrographs N. nucleus, NM nuclear membrane, O nucleolus, G Golgi apparatus, M mitochondria, R reticulum of particle-covered membranes (endoplasmic reticular), P ribosomes.

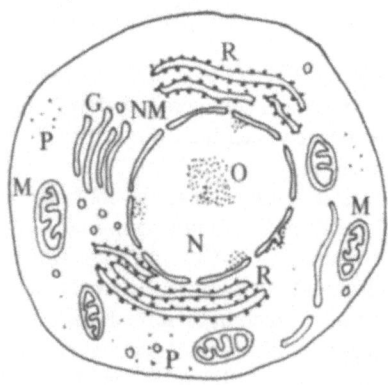

Fig 4 THE EUKARYOTE CELL: ACTIVE AND INACTIVE. The cytology of two forms of cells synthesising proteins: (a) glandular, secreting cell. N nucleus, G Golgi apparatus, S secretory granules. Note numerous particle-covered membranes and polarisation of cytoplasm; (b) retaining cell. M mitochondrion. Fewer Golgi membranes, no particle-covered membranes. Numerous ribosomes.

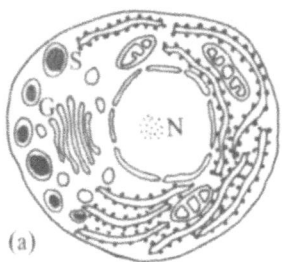 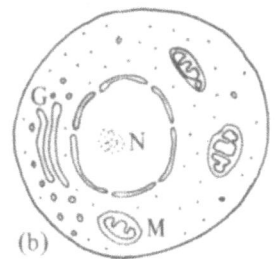

Fig 5 THE EUKARYOTE PLANT CELL. Fine structure of a meristematic (embryonic) plant cell. The nucleus contains a large granular nucleolus n and is enclosed in a double-layered membrane pierced by 'pores'. The Golgi cluster of membranes G is well developed, and several paired membranes D, perhaps early steps in the formation of vacuoles such as V, are shown (m are mitochondria). The cell wall is usually thin at this stage. The cytoplasm is very dense, with ribosomal particles as in embryonic animal cells. PM is plasma membrane.

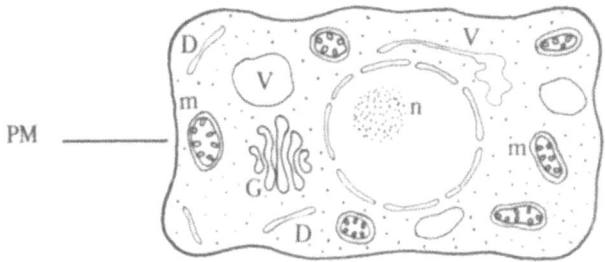

EUKARYOTE CELLS

There are two primary subdivisions of eukaryote cells, nucleus and cytoplasm. The nucleus contains the genetic material and the code, which directs the cell's activities. The cytoplasm organises those activities and, through the mitochondria, directs the cell's energy exchanges. The various organelles present in the cytoplasm are important for these functions and are seen to be more developed and more active in secreting cells than in structural cells.

Compare the much simpler prokaryote cell of the bacterium, which can nevertheless manufacture a wide range of enzymes.

advantageous were retained. Thus, as with life in general and in conformity with the principles of ecology, randomness gave way to order.

The bacterial cell has no nucleus, the DNA being dispersed throughout the cell cytoplasm. There are genes controlling heredity, but these are not organised as chromosomes. Most experts agree that they have no organised mitochondria which form the powerhouse of nucleated organisms. It took another 600 million years for a eukaryote (cell with a nucleus) to be developed from the prokaryotes (cells without a nucleus). The development of the eukaryotes must be regarded as another of the outstanding milestones in earth's history, especially since, when they appeared, many of them contained chlorophyll and were capable of true photosynthesis. However, chlorophyll was not a new invention since it is present in the green and purple bacteria, very primitive organisms which live in estuaries and use sulphur sources for their energy requirements. In addition they use energy acquired from light at the red end of the spectrum, but they do not synthesise carbohydrates from carbon dioxide, nor do they excrete oxygen.[9]

The blue-green algae were the first chlorophyll-containing organisms to appear which could synthesise carbohydrates, evolving oxygen into the atmosphere during the process. Like some fungi, they were coenocytes, that is to say that though they were eukaryotes the nuclei were not separated into individual cells but there were many nuclei in the cytoplasm of one cell envelope. The chlorophyll too was distributed throughout the cell sap and not confined to chloroplasts or plastids, as in more highly evolved organisms. Some, such as *Nostoc,* can fix atmospheric nitrogen. The appearance of the blue-green algae was followed by that of the unicellular chlorophyllous algae, in which the chlorophyll is present in organised plastids. In was from the organisation of these one-celled algae into colonies, with eventual differentiation of function amongst the cells, that the metazoa or multi-cellular life forms developed. To look at the time scale again:

	(Years ago)
Origin of the earth	4,500 — 5,000 million
Bacteria	3,300 million
Blue-green algae	2,700 million
Fungi	2,300 million
Metazoa	600 million

It has, thus, taken the greater part of earth's history, 4,000 of say 4,600 million years, to produce a simple animal like a jellyfish.

Before passing to the pollution of earth's atmosphere with oxygen, let us take a quick look at the great evolutionary step resulting in the appearance of the eukaryote cell. The first eukaryotes to appear in the earth's record were the blue-green algae and the fungi, both coenocytes though some fungi and some stages of fungal development are unicellular. The fungi, like the bacteria, live saprophytically and use a multiplicity of chemical processes to obtain their energy requirements. The unicellular organisms, protozoa and protophyta, followed, and with them was developed a sophisticated type of cell, in all respects comparable with the cells that are the basis of metazoan life from worm to man. Each cell contains a complete metabolic system, is equipped for locomotion, to detect and seek food, for ingestion absorption and excretion, for sexual and asexual reproduction, for gene inheritance, and for power utilisation. It is sometimes suggested that these cells, and all cells of metazoa, started as hybrids in the sense that they are not uniform structures. This idea started with the plastid, the structure which contains chlorophyll and closely resembles some free-living protozoa. It is supposed that the plastid may have originated from a protozoan which came to live in symbiosis with a non-chlorophyll-bearing protozoan of another type, and gave it protection and powers of mobility. Such an idea has support, because in the early days of eukaryote evolution before oxygen was plentiful in the atmosphere, symbiosis became a feature of a number of primitive organisms, such as sea anemones, many of which harbour green flagellates in their tissues. Secondly, when cell division takes place, the plastid divides independently of the host cell. Thirdly, plastids have been removed from the host cells and can for a certain time live independently of them, carrying on their work of photosynthesis.

It has also been suggested that the cell and its nucleus may originally have been separate entities which came together fortuitously. For instance, the nucleus of one cell can be removed and transplanted into another cell similarly deprived of its nucleus; it can take control and regulate the functions of the cell into which it is transplanted. These are interesting speculations for which there can be no proof, but which could explain the ways in which the eukaryotes were evolved. In any case, primitive eukaryotes have as diverse methods of acquiring and using energy as have more primitive life forms. It was not until much later that energy use became standardised in oxidation/reduction processes; indeed this may have been a process forced on living systems outside specialised habitats by the increasing pollution of the atmosphere with oxygen.

Once the blue-green algae and the unicellular chlorophyll-bearing protozoa appeared on earth, a new driving force of life came into existence dependent on photosynthesis, that is the trapping of the sun's protons on chlorophyll and use of the energy to convert carbon dioxide into carbohydrates and fat in which the energy is stored in chemical form for recovery when required. It suffices here to say that the raw material of the chemical process, carbon dioxide, is acquired from the atmosphere, and oxygen is the end product which is excreted into the atmosphere. The earlier life forms, as we have seen, required pre-existing molecules of organic materials or biochemicals on which to live. These would become scarcer as some screening of the earth's surface occurred by oxygen and ozone derived from photo-dissociation of water and plutonic rocks, and in any case would not occur at any depth in water. It is unlikely that saprophytic life could have continued indefinitely unless a new energy source could be tapped. Such now became available, but owing to the poisoning of the atmosphere by oxygen it would also have become self-limiting unless new metabolic mechanisms had been evolved, to which the presence of oxygen would be inimical. Furthermore, when the oxygen pressure reached 1.0 per cent of present levels — known as the Pasteur point — oxidation/reduction systems of energy use become more efficient than fermentation. This point was reached at the beginning of the Cambrian Period some 550 million years ago, when a great population explosion of varied life forms occurred. This was another of the great landmarks in earth history. Atmospheric oxygen must have been built up steadily from then on, until the stable condition of earth's environment of today was reached. In time the ozone belt was lifted to a sufficient height to enable colonisation of the land as well as the sea, and the age of lichens, mosses, giant ferns, and gymnosperms led into the Carboniferous era responsible for our fossil fuels.

Since that time, the earth has passed through many vicissitudes. Some aspects of this history we shall study in greater detail in later chapters. The great and prolonged age of reptiles came and went. The primitive and originally stupid warm-blooded mammals became dominant and radiated into the world's habitats, evolving into many forms, extinct and extant today. The whole proceeded according to the dictates of ecology, until man came to defy ecology some 10,000 years ago, and that must be regarded as the last of our turning points in earth history. As a final exercise in this chapter, we can list earth's turning points:

	(Years ago)
Birth of the earth	4,500 – 5,000 million
Appearance of first life (bacteria)	3,300 million
Photosynthesis	2,700 million
Eukaryotes (fungi)	2,300 – 1,000 million
Metazoa	600 million
Pasteur point	550 million
Agricultural revolution of man	10,000 million

In nearly 5,000 million years of earth's history, there have been but six revolutionary turning points. Man's interference with natural ecological processes, only 10,000 years ago, has been one of them.

NOTES

1 L.V. Berkner and L.C. Marshall, 'History of Major Atmospheric Components', *Proc. Nat. Acad. Sci.* vol. 5, 53, pp. 1215-26 (1965)
2 E.C. Olsson, 'Climatic Change and its Influence on Life and Habitat', in R.N.T -W - Fiennes, *'Biology of Nutrition'*, Pergamon Press, Oxford, 1972
3 Berkner and Marshall, op. cit.
4 For a review of chemical evolution, see Crawford, in Fiennes, op. cit.
5 For reviews, see Berkner and Marshall, op. cit., and L.V. Berkner, 'On the Origin and Rise of Oxygen Concentration on the Earth's Atmosphere', *J. Atmos. Sci.,* 23, pp.133-43 (1965)
6 Berkner and Marshall, op. cit
7 N.W. Pirie, 'Characteristics of Living Things', in Fiennes, op. cit.
8 Crawford, op. cit.
9 See E.J. Ferguson-Wood, 'The Food Cycle in Marine Environments: Marine Algae', in Fiennes, op. cit.

Further Reading

Bernal, J.D., *The Origin of Life,* Weidenfeld & Nicolson, London, 1967
Fiennes, R.N.T -W - , *Biology of Nutrition,* Pergamon Press, Oxford, 1972
Fox, S.W., *The Origin of Prebiological Systems,* Academic Press, N.Y., 1965
Kulp, J.L., 'Geological Time Scale', *Science,* 133, pp.1105-14 (1961)
National Academy of Sciences Symposium, 'The Evolution of the Earth's Atmosphere', *Proc. Nat. Acad. Sci.* (Wash.), 53, p.1169
Oparin, A., *The Origin of Life on Earth,* Oliver & Boyd, Edinburgh, 3rd ed., 1957
Oparin, A. (ed.) *The Origin of Life on Earth,* Pergamon Press, Oxford, 1964
Oparin, A., *Life: Its Nature, Origin, and Development,* Oliver & Boyd, Edinburgh, 1961
Stanton Hicks, Sir C., 'The Nutritional Requirements of Living Things', in Fiennes, op.cit.

3 PRIMITIVE LIFE FORMS AND THE EVOLUTION OF METAZOA

The metazoa, which we regard as the higher forms of life, could not exist without the lower forms evolved in the early days of earth history. The lower forms, on the other hand, could, and for 2,700 million years did, exist without the higher. Furthermore, their biomass, their total weight of living material, vastly exceeds the total biomass of the vertebrate kingdom. It is well, therefore, to devote a chapter to their study and to examine their activities.

Whereas the metazoa, whether plant or animal, employ a single energy system, the reduction/oxidation or 'redox' system, lower organisms employ a variety of methods to obtain and use their energy. At this point, it will be helpful to introduce a few terms that are used to describe methods of nutrition. Living things are either 'autotrophs' or 'heterotrophs'. *Autotrophs* use inorganic nutrients, and fix carbon dioxide or use carbonate or sodium bicarbonate as their source of carbon. This group thus includes the plant kingdom. *Heterotrophs* require organic compounds both for energy metabolism and carbon assimilation. The group thus includes the animal kingdom. Living forms which use stored resources for their energy requirements, as do the metazoa, are described as *endergonic;* those that use external energy sources are known as *exergonic.* Yeasts are typical examples of exergonic organisms, since they obtain their energy requirements by degrading glucose, present in their environment, into alcohols. *Chemotrophs* are organisms obtaining energy by the oxidation of chemical compounds derived from the environment (exergonic). Chemotrophs are of two groups: *chemolithotrophs,* which use oxidisable inorganic substances, such as molecular sulphur or hydrogen sulphide or ammonia; and *chemo-organotrophs*, which use only organic substances for the same purpose.

Phototrophs obtain their energy from radiant energy. They contain chlorophyll and are able to trap and use the photons of incident light.

Photo-lithotrophs use a combination of radiant energy and oxidisable inorganic compounds.

Photo-organotrophs use a combination of radiant energy and oxidisable organic compounds.

Two other terms are worthy of mention: *phagotrophy* or *phagocytosis,* which is the ingestion by single cells of other cells or particles; and *pinocytosis,* the ingestion by individual cells or droplets.

Within these definitions, there exists an enormously complex variety of mechanisms for feeding and energy use. The higher animals of today

Fig 6 ENERGY PATHS OF AN ECOSYSTEM 1. Fate of energy incorporated by autotrophs in Cedar Bog Lake, Minnesota in gram calories per square centimetre per year.

Data of R Lindeman. 1942. Ecology 23: 399-418

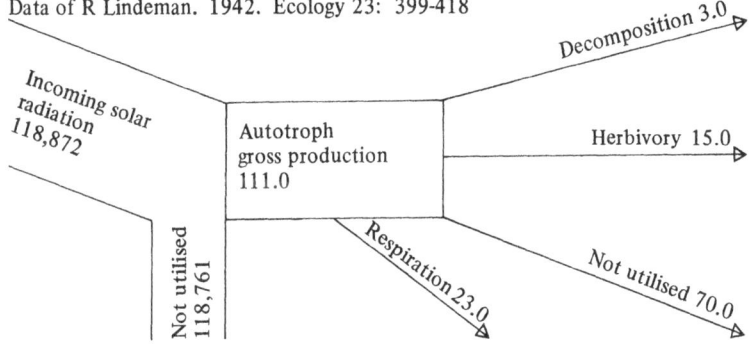

Fig 7 ENERGY PATHS OF AN ECOSYSTEM 2. Fate of energy incorporated by herbivores in Cedar Bog Lake, Minnesota in gram calories per square centimetre per year.

Data of R Lindeman. 1942. Ecology 23: 399-418

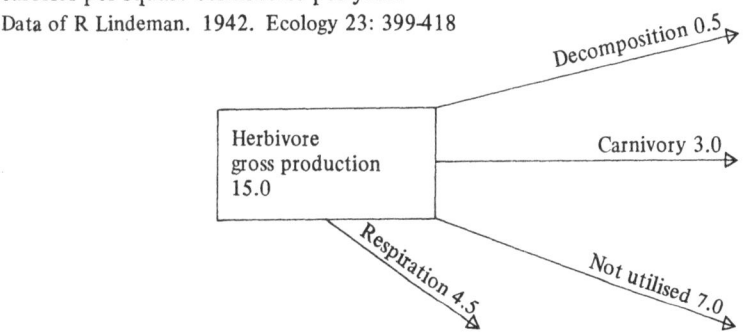

Fig 8 ENERGY PATHS OF AN ECOSYSTEM 3. Fate of energy incorporated by carnivores in Cedar Bog Lake, Minnesota in gram calories per square centimetre per year.

Data of R Lindeman. 1942. Ecology 23: 399-418

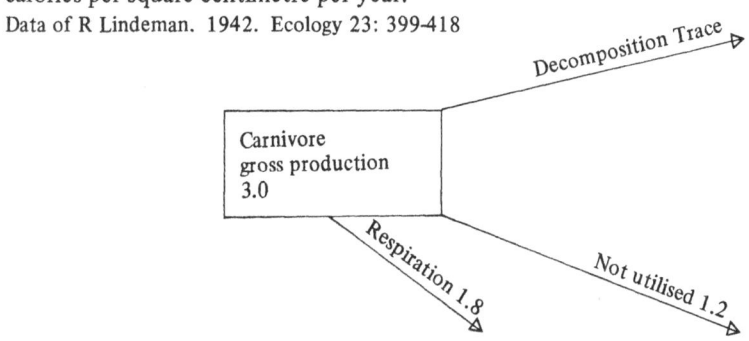

Fig 9 ENERGY PATHS OF AN ECOSYSTEM 4

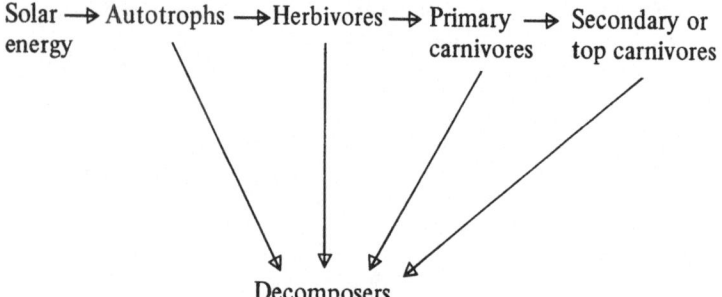

have virtually lost the power to digest vegetable matter; they are unable
on their own account to split the molecules of cellulose and other
higher polysaccharides. Thus, herbivorous animals possess large
fermentation chambers, in which micro-organisms do the job for them.
Furthermore, higher animals are no longer able to synthesise some of
the vitamins essential to them, and this too is done for them by
bowel-living micro-organisms.

Micro-organisms

However, we must now study the various groups of micro-organisms that
are essential components of ecosystems. These are the bacteria, the
fungi, the algae, and the protozoa.[1]

The bacteria, as we have seen, are supposed to have been the earliest
forms of true life to have been evolved. These organisms possess a
multitude of ways in which they obtain their energy. They are, of
course, responsible for many infectious diseases, but their activities are
for the most part beneficial and our kind of life could not continue
without them.

The green and purple sulphur bacteria are the only bacteria to
contain chlorophyll, as we have already seen. They are found in
habitats where few other life forms can exist, that is among the rotting
detritus of sea grasses in estuaries. The decomposition of the detritus
produces more or less anaerobic conditions, that is absence of oxygen,
and great acidity. For this reason, many organic chemicals are present
in reduced (more complex) form instead of being oxidised; there is
methane (marsh gas) in place of carbon dioxide, ammonia instead of
nitrate or nitrite, and sulphide instead of sulphates. There are also a
great many unsatisified iron compounds, often linked with sulphur.

32

Stagnant lagoons become black with iron compounds and the smell of hydrogen sulphide becomes overpowering. Such conditions are suitable for the green and purple sulphur bacteria, which are virtually unique in being able to live in such a situation. These bacteria are photosynthetic, but obtain their energy by knocking electrons off H_2S (hydrogen sulphide) instead of H_2O (water), used by other photosynthetic plants; this is the reason why their photosynthetic activities do not release oxygen. The ecology of such a habitat depends for its initiation on the photosynthetic activities of these bacteria.

When conditions become less extreme, some other bacteria continue the process. Ciliates (protozoa) prey on them, and even some nematode worms, crustaceans and molluscs become established. The sea grass flats which then appear are very unproductive habitats, but with understanding of the ecology and proper treatment, large harvests of fish protein can be produced by culture methods. The coloured bacteria survive beneath the other vegetation, because they only use light toward the red end of the spectrum. They can thus 'scavenge' the more penetrating wavelengths, which have filtered through carpets of sea grasses.

Some chemolithotrophs (sulphur bacteria) also use sulphur and are abnormally resistant to acid media. One indeed can tolerate strong sulphuric acid, and some of them can cause erosion of concrete and alkaline rocks. The important soil-nitrifying bacteria belong to this group, converting ammonia to nitrite and then nitrate ($NH_4 \rightarrow NO_2 \rightarrow NO_3$). Another group uses molecular hydrogen as an energy source, and another group iron compounds. The latter occur in iron-containing soils such as bogs and swamps, where they produce rust-coloured patches. Where present in some forest soils, they may precipitate iron below the topsoil causing an iron pan to develop and converting the forest in time into bog.

Most true bacteria (non-chlorophyllous) and all the fungi are chemo-organotrophs. Metabolically, bacteria and fungi are rather similar and the fungi probably evolved from the bacteria. Although the bacteria are mostly unicellular, there are important filamentous forms which appear to be intermediate, such as *Actinomyces*. However, all fungi have primitive eukaryote nuclei, whereas the filamentous bacteria do not; nor do bacteria have distinct mitochondria, which fungi have. The more primitive fungi such as the yeasts have rather primitive nuclei and they appear to be the earliest representatives of eukaryotes. Both bacteria and fungi are variable in their requirements of oxygen. Some can only live in an atmosphere containing oxygen; some can only live in an atmosphere devoid of oxygen; yet others require diminished oxygen; and others finally can live either in an aerobic or an anaerobic atmosphere. Their metabolic pathways can be oxidative or fermentative.

The oxidative reaction requires molecular oxygen as the final electron acceptor. The fermentative reaction is an incomplete one in that molecular oxygen is not required and energy requirements are obtained from organic chemicals only degraded part-way to products of lower molecular weight. In both cases, the energy is transferred to the substance ATP, described later.

The range of materials that can be attacked by the chemo-organotrophic bacteria covers virtually any organic materials found in nature. It is, however, an anomaly, revealing of the hit-and-miss nature of chemical evolution, that there is a great dearth of organisms capable of attacking the man-made synthetic organic chemicals, whether the organo-chlorine agricultural chemicals or the plastics, such as polythene and polystyrene. For this reason, the man-made synthetics have become a serious problem of waste disposal and pollution. This immunity from attack, in another way, has its uses because these substances are not only spared microbial action, but are not even attacked by the defence cells and fluids of the vertebrate body. They can, therefore, be used for organ implants, such as heart valves, or teeth, and will not be resorbed.

Amongst the hydrocarbons, there is one bacterial genus, *Methanonomas*, which can oxidise methane, and many can to a certain extent use ethane, propane and butane. Some yeasts too can attack a range of such hydrocarbons. There are even organisms capable of attacking fractions of crude oil, and it has been suggested that by their use food products could be synthesised from crude oil.

There exist species of bacteria and fungi which can split virtually all the higher polysaccharides of high molecular weight, some of them producing intermediate products, such as dextran, which are of commercial value. Organisms with such properties are also responsible for 'dry rot' of timber, 'soft rot' of vegetables and other deleterious processes. Organic acids are commonly released during the fermentation of carbohydrates; acetic acid is made by the vinegar bacteria; propionic acid important in the manufacture of soft cheeses is made by *Propionibacterium;* and lactic acid, responsible for clotting of milk, is made by lactobacillus. Some organic acids are themselves utilised as energy sources; such are citrate, tartrate, malonate and lactate.

Other organisms obtain their energy and carbon from breakdown of amino acids, by removal of the amino group. Others produce enyzmes capable of attacking nucleic acids. Fats too are hydrolysed by some groups of bacteria to yield glycerol and fatty acids. There are even bacteria which can degrade chitin, the material of the hard shells of lobsters and insects.

Bacteria and fungi then are the great scavengers of the world. Man-made synthetics apart, there are few organic materials they cannot decompose. In this, they play a vital part in the economy of nature.

We must revert now to the blue-green algae already noted as the earliest true plants in which the full photosynthetic process had been developed. It was due to their influence that the atmosphere first became seriously contaminated with oxygen. Structurally, the blue-green algae are primitive and show greater affinities with the green bacteria than with the true algae. Like the bacteria, they do not have a defined nucleus within a nuclear membrane; their chlorophyll is dispersed throughout the cell body, not in defined chloroplasts or plastids; and like the bacteria, they do not show the streaming of cytoplasm characteristic of higher forms, such as fungi. On the other hand, they are mostly filamentous; they possess chlorophylls resembling those of true algae and higher plants; and they use carbon dioxide and water in the process of photosynthesis instead of hydrogen sulphide, used by the green bacteria. Like fungi, their reserve food is glycogen (animal starch), not starch. The filaments of the blue-green algae are embedded in a tough mucilaginous sheath, so that the colony becomes a slimy green layer often seen in river estuaries at low tide. Underneath the slime are often also found the green bacteria. They may also be free-floating in water or have filaments radiating into the water. A large group also lives in soil habitats and these have the power, rare in nature, of fixing atmospheric nitrogen as well as using light as an energy source. Some, protected by their mucilage, can even creep over hard substrates. Thus their range of habitats, which are always primitive, is wide, and they have been found both in arctic waters and in hot springs. Some blue-green algae can taint water supplies or render them toxic to animals.

Perhaps the chief importance of the blue-green algae is in maintaining the aeration of estuary waters, a role which is vitiated when there is estuary pollution by sewage; this stimulates the growth of a surplus of oxygen-using micro-organisms, which quickly deplete the waters of oxygen and make it impossible for fish and other forms of aquatic life to survive. As pioneers, some blue-green algae live symbiotically with fungi as lichens. In this role, both their symbiotic and their nitrogen-fixing properties would be of value to the fungus, and the fungus, able to absorb moisture from the atmosphere, would protect the algae from desiccation.

Metabolically, the blue-green algae have taken a great step forward over the green bacteria, in that they have developed a form of true photosynthesis in which water replaces hydrogen sulphide in energy conversion and oxygen is given off into the atmosphere. This revolution paved the way for the evolution of higher life forms. Nevertheless, they still retain an important place in earth's ecology, both in water and on land.

Unicellular eukaryote plants do not appear in the fossil record until some thousand million years ago, although eukaryote fungi were in existence some two thousand million years ago. It cannot be said, therefore, whether the unicellular plants were developed from fungi which acquired chlorophyll, or from blue-green algae which independently developed a eukaryote nucleus, or from some other form of life of which we have no record. The material, and the sources from which it is obtained, are so scanty and in other ways unsatisfactory, that any conclusions based on them must be largely speculative. Suffice it to say that further evolution, the development of the metazoa and higher life forms, depended entirely on the eukaryote cell. The chief feature of this cell is the specialisation of function within the cell, which is seen in three basic ways. First, the genetic chromatin is packed into a nucleus within a definite nuclear membrane. Secondly, the power systems of the cell are packed into definite organelles, known as mitochondria. Thirdly, the cell chlorophyll is packed into an organ of increasing complexity known as the plastid or chloroplast. Organs of ingestion, excretion, food digestion and storage, and locomotion also appear amongst some single-celled eukaryotes. At the same time, genetic material has become aligned on chromosomes and mating cells undergo reduction division (meiosis) preparing them for fusion.

Since the appearance of both fungal and plant eukaryotes, the fungi have gone their own way, retaining their primitive methods of obtaining energy, and have not contributed to the main evolutionary stream. It is, therefore, to the plant eukaryotes that we must look for further development. Many of these forms, especially the green flagellates, have the ability to live either as autotrophs or as heterotrophs; they can live by photosynthesis alone, or by assimilation of organic materials; alternatively some photosynthetic organisms require organic material as well. In addition, there exist protozoa, otherwise identical with those that are photosynthetic but which have no chloroplast. If some photosynthetic protozoa are kept in the dark with a sufficiency of organic material in the medium, they will lose their chlorophyll but regain it when again exposed to light. Amongst these primitive organisms one cannot therefore distinguish between plants and animals. However, once evolved these eukaryotes developed a balanced ecological system of their own, consisting of plants, plant-living organisms, grazers, predators and scavengers.

Evolution of Metazoa

The beginnings of our modern world came with the development of metazoan life some 600 million years ago, and the great explosion of life forms in the seas with the beginning of the Cambrian period some 550 million years ago. At that time, the foundation stock of all our

37 modern life forms had appeared. Once started, therefore, the colonisation of the earth's habitable surfaces proceeded with comparative speed. The evolution of the metazoa was initially a somewhat simple process, and can be traced in colonial forms that still exist today. In these, the individual flagellate cells are arranged in a sphere with the flagella facing outwards, the whole colony protected by mucilage, as with *Volvox; Pandorina* has the form of a solid sphere, and *Gonium* a flat plate. In some, fine protoplasmic strands connect the cells. They are found in fresh, marine, and brackish waters. Amongst them, division of labour is to be found in the more advanced, in that only certain cells reproduce while the remainder are in a vegetative condition. By a simple involution of a form such as this, we reach an organism remarkably similar to the planktonic larvae of marine worms, jellyfishes, starfishes, sea anemones, and the Chordata – ancestral to the vertebrates. What regulates the differentiation of cells, originally homogeneous, into different tissues, is still not understood. However, we need not dwell on this in a work on ecology. Let us see how the organisms live and what part they play in the earth's economy.

When we think of the protozoa, we have to realise that we are dealing with living cells which possess basically the same limitations as the cells of metazoa, the cells of our own bodies. The myriad chemical systems used by bacteria and fungi are not available to them. The nutrients they require are in many respects similar to those we need. If we cannot synthesise a certain vitamin, the chances are that protozoa cannot do so either. Their general energy pathways are similar to our own. They are thus not important in ecology in processes of decay. Their main importance is probably in the seas.

On land, the vegetation is mainly rooted in the soil in the form of grasses, trees, shrubs and other plants. In the oceans, this is only possible in limited habitats, where the water is shallow and where light can penetrate in sufficient amounts to be trapped and used. Consequently, by far the greatest part of marine vegetation is in the form of photosynthetic algae and phytoplankton, which are floating in the upper layers of the water – above 300 metres. Of this, there is a tremendous biomass, whose photosynthetic activities exceed those of terrestrial plants. Thus a great part of the earth's atmospheric oxygen is returned to the atmosphere through the agency of these organisms, comprising mostly the diatoms, and the green flagellates.

This vegetation is grazed by other micro-organisms, which are the primary predators of the food chain or web. The process is similar to that on land, although only one-celled organisms are involved, except that the growth is so prolific as to be used for food even by the largest animals on earth, the baleen or rorqual whales.

Fig 10 SIMPLE COLONIAL PROTOZOON WITH LITTLE
ORGANISATION. *Conium pectorale:* colony of sixteen individuals,
each with two flagella (x c480). (a) surface view, (b) side view, N nuclei,
cv contractile vacuole, st stigmata.
From Minchin. 1912. After Stein

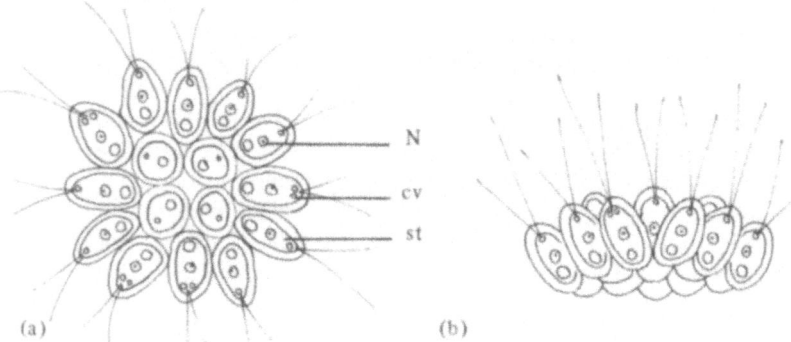

Fig 11 SINGLE-CELLED PROTOZOA FROM WHICH COLONIAL
PROTOZOA ARE DERIVED

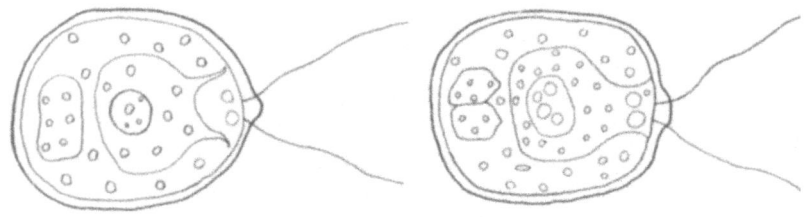

Fig 12 STRUCTURE OF A COLONIAL PROTOZOON. The structure
of *Volvox*, showing the different arrangement of the cells in the two
species *Volvox globator* and *Volvox aureus*.

After Janet and Klein

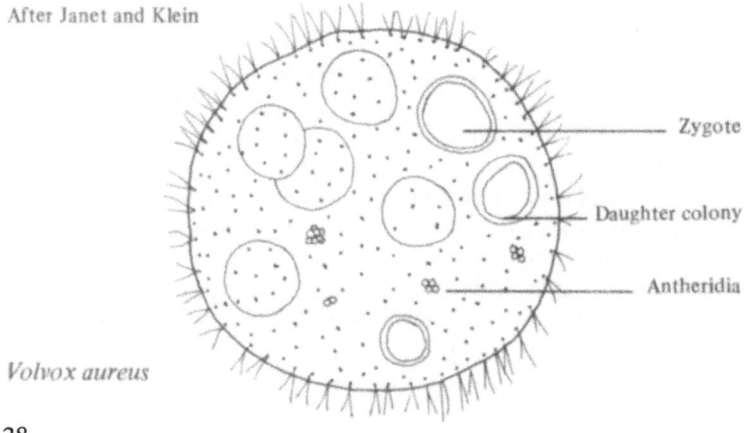

Volvox aureus

Fig 13

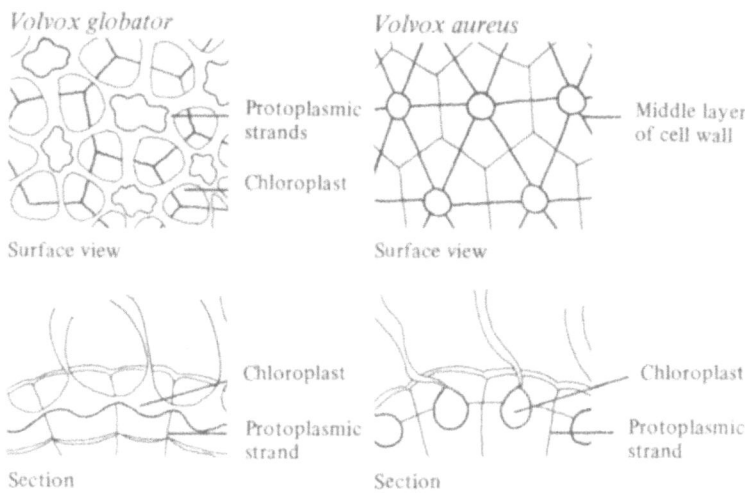

Volvox globator

Protoplasmic strands

Chloroplast

Surface view

Volvox aureus

Middle layer of cell wall

Surface view

Chloroplast

Protoplasmic strand

Section

Chloroplast

Protoplasmic strand

Section

Fig 14 DEVELOPMENT OF METAZOAN LARVAE. Cleavage and gastrulation. There are three types of gastrulation shown here: invagination in *Littorina,* epiboly in *Crepidula,* and ingression in *Patella.*

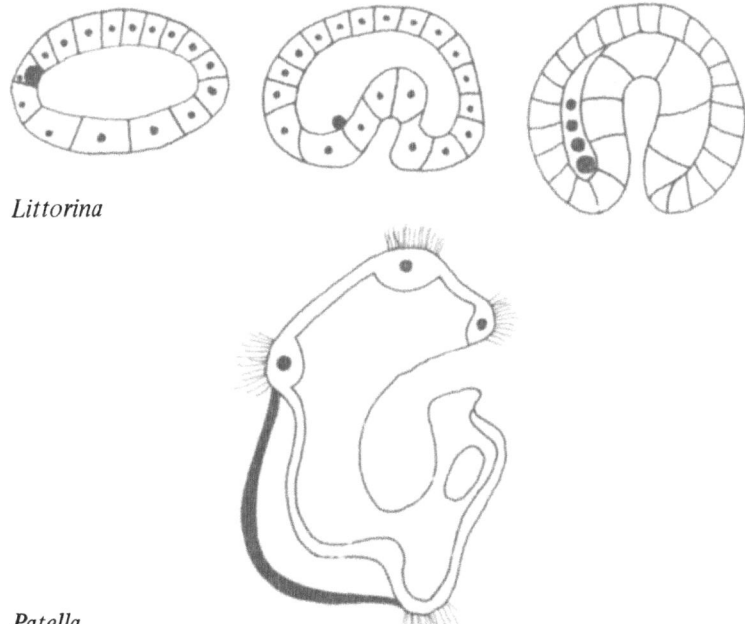

Littorina

Patella

Thus the aquatic food chains or webs, except where waters are shallow, start with these microscopic organisms, which ultimately support the fish and other populations of the seas.

On land, green algae and protozoa are abundant in the scum of ponds, ditches, and stagnant waters, playing a part in the ecology of such habitats. Amongst the predatory protozoa are colourless flagellates, ciliates, and amoebae. These are present in soils in large numbers, where they prey on the bacteria. They in their turn will provide food for earthworms, predaceous fungi, and other soil-living fauna. In terrestrial habitats, nevertheless, protozoa and algae are of comparatively minor importance, except for their activities as symbiotes and parasites of plants and animals.

Every great group of protozoa has produced parasitic forms which are of medical importance. From the flagellates have arisen the leishmanias and the trypanosomes, which cause diseases such as kala-azar and sleeping sickness; some forms of amoebae cause amoebic dysentery; some ciliates are serious bowel pathogens. Finally, there is a whole class, the sporozoa, no members of which have any other way of life than the parasitic. Amongst the sporozoa are the malaria parasites, which are responsible for more human deaths than any other organism of disease.

In summary, the protozoa exploited to the full the opportunities open to the unicellular eukaryote way of life. Their main importance lies in providing the vegetation of the seas, and without them life in the seas would be far less prolific than it is; in doing this, they serve to stabilise the oxygen level of the atmosphere. On land, they have taken largely to a parasitic mode of life and are amongst the most important agents of disease in higher animals. The main line of advance in the eukaryote world lay with the metazoa, for which the eukaryote cell is splendidly adapted.

Following the Cambrian period[2] colonisation of the seas was rapid: first the invertebrates, followed by the vertebrates as cartilaginous fishes, and later re-invaders from the land or fresh water, reptiles, birds and bony fishes. The heyday of the fishes started in the Silurian and the Devonian, some 450 to 400 million years ago, the earliest forms being cartilaginous fishes like the sharks. During the early Devonian, the first traces of life on land appear in the form of vascular plants and a few invertebrates. Vertebrates first appeared on land at the end of the Devonian, 400 to 350 million years ago; there developed successively amphibians and reptiles, then mammals and birds.

The great age of reptiles lasted 450 million years. Its sudden and drastic end is widely attributed to climatic changes, which rendered them powerless in competition with the better adapted early mammals. However, this supposition is hard to support. There is no satisfactory evidence that a sudden climatic change occurred at the end of the Mesozoic, and mammals at that time were small, insignificant little creatures, which could certainly not have been in competition with the reptiles. Furthermore, this theory does not explain why the aquatic reptiles should have disappeared at the same time as the terrestrial.

The disappearance of the reptiles is one of the great mysteries of earth history. However, it happened some 100 million years ago and the radiation and expansion of the mammals into the vacant ecological niches was in full swing in the Eocene epoch some 50 million years later. They inherited a world that was temperate and warm, but in which trends were towards increasingly cool climates. Climatic deterioration was accelerated during the latter third of the 50 million years of the Cenozoic era, culminating in the glaciations of the Pleistocene epoch 2.5 million years ago. The causes of the great glaciations of the Pleistocene are not known. There were four major glaciations interrupted by warmer interstadials. The last glacial epoch started to recede some 20,000 years ago. The glaciations had a profound effect on the ecology, both as regards plant and animal life. In the tropics, the glacials were represented by pluvials with excessive rainfall, but even at the Equator the glacier limits descended to 5,000 feet. Even at the beginning of the human era, the Sahara was mostly fertile and inhabited by nomadic peoples who hunted and kept cattle. Since then, increasing desiccation has been accelerated by human malpractice. Man himself not only survived the glaciations, but they appear to have suited his way of life as hunter and predator in small bands. Wildlife was abundant, following the ice fringes north in Summer and south in Winter; meat supplies were easily won by hunting or trapping. During the winter, the meat could be deep frozen in ice or snow, so there were no shortages. Animal furs and skins provided warm clothing and the marvellous invention of fire kept the caves or mammoth-tusk hutments warm and snug.

Man's way of life was overthrown when the climate became warmer in the Mesolithic era, for the forests spread over the land. Hunting became more difficult and less rewarding and changes of diet were necessary for survival. Then came the agricultural revolution with the first hesitant attempts to till the land and grow grain and domesticate stock.

The profound effects of climatic change, then, were something that had not been experienced since the early Cambrian nearly 550 million years previously. It is now 10,000 years since the last glaciation receded. Are we enjoying an interstadial after which the ice will return? Or are we entering another prolonged period of temperate climate? If the ice is to return, when will this happen? To these questions, we know no answer. A return of the ice to its previous southern limits along the line of the Thames and south of the Alps would render life as we know it impossible, and the weight of the ice on land surfaces would tilt them and cause widespread flooding.

Meanwhile, it behoves us to study and overcome our problems as we find them, but with the thought that nature is more powerful than we are and will inevitably have the last word.

NOTES

1 See J.E. Smith, 'The Nutrition of Bacteria and Micro-Fungi', in R.N.T -W- Fiennes, *Biology of Nutrition,* Pergamon Press, Oxford, 1972, and S.H. Hutner, H. Baker, O. Frank and D. Cox in Fiennes, op. cit.
2 See E.C. Olsson, 'Climatic Change and its Influence on the Evolutionary Development of Life Forms', in Fiennes, op. cit.

Further Reading

Borradaile, L.A., and Potts, F.A., *The Invertebrata,* University Press, Cambridge, 1958

Clark, R.B., *Dynamics in Metazoan Evolution,* Clarendon Press, Oxford, 1964

Cloudsley-Thompson, J., *The Temperature and Water Relations of Reptiles,* Merrow Publishing Co., London, 1972

Mercer, E.H., *Cells and Cell Structure,* Hutchinson Educational, London, 1961

4 PHOTOSYNTHESIS AND THE ENERGY PATHS
OF LIFE

In the preceding pages, continual reference has been made to energy and its modes of use in living things. It will be well now to pause for a consideration of the mechanics of life and its power sources. The system is often spoken of as 'energy flow'. Energy is trapped into the system and made to flow upwards into energy stores, from which it is released at need to provide the driving force of metabolic processes. The energy can be studied in relation to a single plant or animal, or groups of plants or animals; that is, energy flows can be traced through a habitat and its 'associes', through the primary producers, predators and scavengers back to the products of decay and recycling of the elements.[1]

In the preceding pages also, reference has been made to a property of living systems, by which increasing order emerges from disorder. Habitats under the dictates of ecological laws become ordered; living systems were initially varied, but have become stereotyped; the probiotic earth was disordered but has acquired increasing order; the atmosphere has developed a constant composition. The Second Law of Thermodynamics decrees that natural processes will flow towards an increase of 'entropy' or randomness until equilibrium is reached; flows are downhill, not uphill. A hot body will yield heat to a cold, until the two reach the same temperature. A high pressure system will yield pressure to a low pressure system. These principles regulate equilibrium in meteorological conditions and between oceans and atmosphere. In biological systems, however, energy flows move towards greater and more complex organisation against the dictates of the Second Law. This can only occur because of the acquisition and use of energy acquired from natural sources, the chief of these being the radiant energy of the sun.

We have seen how the shortwave radiation of the sun in probiotic days could have, and no doubt did, mediate the synthesis of 'biochemicals' of high molecular weight. The most primitive of living organisms are those which can only obtain energy by degrading biochemicals – in some case chemicals – thereby releasing and using the energy that had been stored in them. This process would be self-limiting, because the ultra-violet light by which they were synthesised would become less available as oxygen was released and shaded the earth's surface; the ultra-violet light could not have penetrated more than a few metres through water. The development of chlorophyll by living systems led to two important changes: first,

Fig 15 THE FATE OF SOLAR RADIATION. Energy intake at the earth's surface at midday.

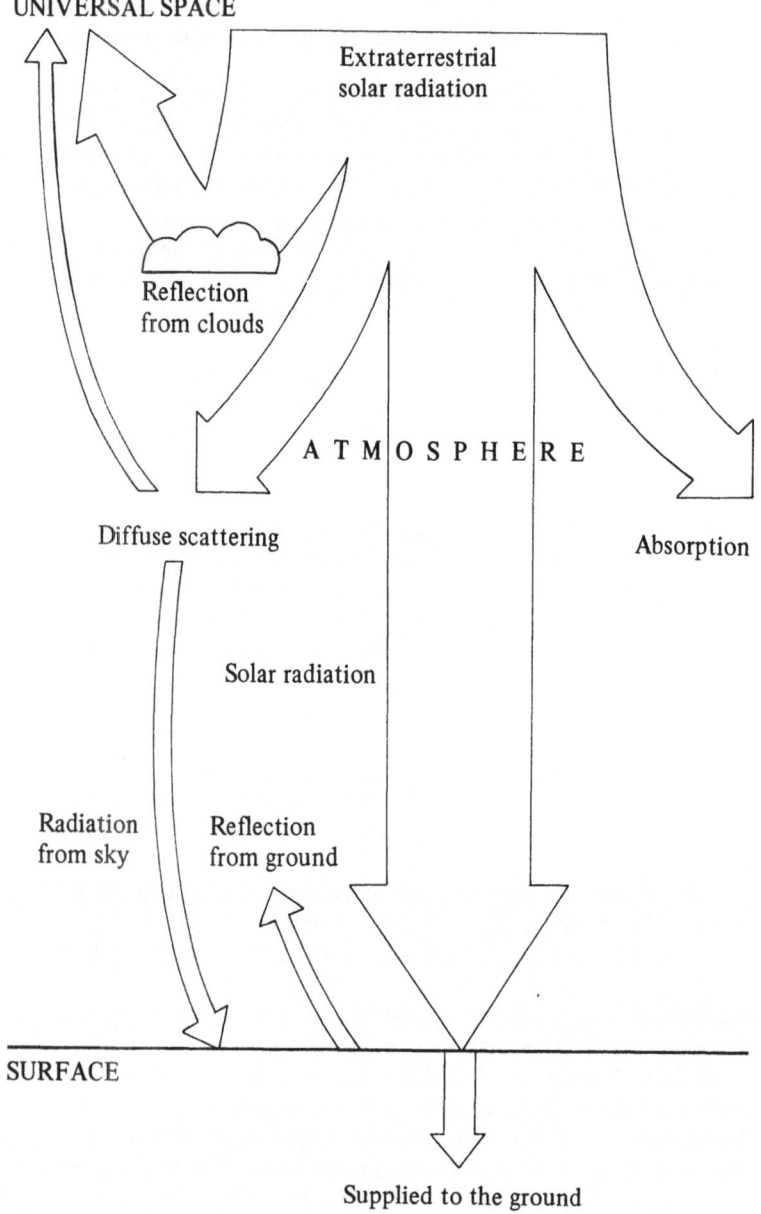

UNIVERSAL SPACE

Extraterrestrial
solar radiation

Reflection
from clouds

ATMOSPHERE

Diffuse scattering

Absorption

Solar radiation

Radiation
from sky

Reflection
from ground

SURFACE

Supplied to the ground

use was made of the visible light rays as the prime energy source; secondly, the trapped energy (electrons) was made to flow into energy stores, from which it could be recovered and converted from chemical energy back into work energy. In biological systems, energy occurs in four forms: (1) chemical or stored energy; (2) electric energy as in nerve transmissions; (3) muscular energy; (4) osmotic energy, as in glandular secretions. A biological system, therefore, must be able not only to acquire energy, but also to direct it at need to any of these four forms. While the basic principle of photosynthesis and energy flow is simple enough, the mechanics are somewhat complex.

So when the earth's surface became screened from the shortwave ultra-violet light of the sun, life would not continue unless an alternative power source were developed; for this, those wavelengths which still reached the surface must be used. The only suitable radiation for the purpose was that of the visible spectrum, and this involved many complexities. However, a system capable of trapping light quanta (photons) was developed, first in the green bacteria, and then further developed in the chlorophyllous protozoa and in the green plants. This development led to the increase in atmospheric oxygen, which enabled the evolution of metabolic systems based on oxygen, and so to the appearance of higher life forms as we know them.

Photosynthesis

Matter, as all know, is composed of atoms, and atoms are composed of a 'nucleus' carrying a positive electric charge round which whirl one or more electrons having a negative electric charge. The path of the electron round its nucleus is known as its 'orbital'; the lowest orbital as the a orbital, and higher orbitals are b, c, d and so on. The simplest atom is that of hydrogen, of which the nucleus consists of a single 'proton', around which orbits a single electron in the a position. The electron is prevented from merging with the nucleus by its own kinetic energy, that is the energy of its own motion; it cannot fly away from the nucleus, because of electrostatic forces. However, if an electric field is applied to the atom, the electron can be raised to a higher orbital, theoretically to an infinite distance. This requires an input of energy, and this energy can be recovered if the electron is again dropped back to its lowest orbital. The energy required to remove an electron to infinity is known as the 'ionisation potential' and it can be measured in electron volts: 1 eV/atom = 23.06 k cal/gram atom* which gives us a relationship between electric force and heat. Electrons at higher orbitals require less energy to remove

*In more modern literature, usually converted to S1 units, which would be without meaning to non-scientists — and indeed to most biologists.

them than those at lower orbitals, a point that is important in photosynthesis. The orbitals of electrons are less simple than here stated, but the Bohr model, as outlined, will suffice for our purposes.

To use visible light as a power source requires the sun's photons to be trapped and their energy used to raise electrons to higher orbitals thus creating a field of electric energy. However, since electronic energy is unstable and cannot be conveniently stored, the further problem arose to convert the unstable electric energy into stable chemical energy. The third problem was to reconvert the stored chemical energy into electric, when it was needed to power life processes, that is to steal the light quanta and manipulate them through biological systems. So, when we study the use of energy by living systems, we are in fact tracing the travels of these light quanta not only through an individual cell or system of cells, but also through an entire ecological system, plant → herbivore → predator → secondary predator → systems of decay and disintegration.

When light rays of the red end of the spectrum fall on the green pigment chlorophyll *a,* two electrons in outer orbitals of its magnesium atom are raised to higher orbitals. Unless they were trapped in some way, the electrons would fall back to their original positions in a minute fraction of a second giving a flash of fluorescence. It is as if a heavy ball were shot upwards by means of a spring. Normally, the ball would quickly fall back to its original position. However, a device could be arranged whereby the ball would be trapped at the summit and then made to fall back through a channel having a series of steps. At each step, a lever or switch could be arranged to be activated during its descent, and in this way the energy of its fall, originally derived from the spring, could be converted into the forms of energy important in biological power systems, mechanical, electric, or chemical (osmotic). In photosynthesis, the excited electrons from the chlorophyll's magnesium are diverted to a high energy chemical acceptor system, the substance NADP being converted by addition of an additional hydrogen atom to NADPH. The source of this hydrogen is important, as we shall see.

At this stage, two electrons need to be replaced in the magnesium atom, and a hydrogen atom must be found for the NADPH. In more advanced photosynthetic systems, this is achieved by splitting the water molecule and temporarily detaching electrons from the hydrogen atom. However, let us look first at the more simple system as in the green bacteria. In these, the NADP → NADPH acceptor system is unnecessary. Electrons are detached from sulphur, a ready electron donor; they are transferred by a chemical transport system to the chlorophyll to replace the electrons lost by the magnesium through the action of red light; those electrons then fall back to the sulphur, some of their energy being trapped to convert ADP → ATP. The basic material used by the green bacteria is H_2S (hydrogen sulphide), so that the end point is molecular sulphur. Certain enzyme systems are also involved. The chemical reaction goes like this: $CO_2 + H_2S \rightarrow 2CH_2O + 2S$. The CH_2O is the basic carbohydrate unit

$$H \longrightarrow \overset{|}{\underset{|}{C}} \longrightarrow OH$$

and these units become combined to form sugars. In the end reaction, the energy trapped in the ATP is used to reduce CO_2.

The use of H_2O as the source of electrons and hydrogen presents greater problems, since the red waves of the spectrum lack sufficient energy to split the water molecule. This is done by the action of blue light trapped by chlorophyll b in a second centre. Thus, two photosystems are used, the original one photosystem I, using red light and photosystem II, using blue. By this means, the water molecule is split and free electrons derived from hydrogen are fed into the system to be transported by a chain of chemical electron donors and acceptors, until they reach the chlorophyll a in photosystem I. The result is the liberation of free oxygen into the atmosphere. The chemical reaction goes like this: $CO_2 + H_2O \rightarrow CH_2O + O_2$. Expressed another way: $6CO_2 + 12 H_2O + 673$ calories + chlorophyll would produce $C_6H_{12}O_6$ (6 carbon sugar) $+ 6O_2 + 6H_2O$. Meanwhile, the hydrogen protons pick up the electrons from the chlorophyll magnesium and enter into the reaction NADP → NADPH. NADPH in its turn reduces CO_2 to carbohydrate.

The only actual input of energy into the system is that derived from the red light in its action on chlorophyll a, although blue light plays a part. No pigment, other than chlorophyll a, reacts to light rays in this way; therefore, all life dependent on oxygen depends on this one pigment. In actual fact, little light of the visible spectrum is wasted by the chloroplast system. The chloroplast is a kind of cell within a cell, and contains numerous 'grana', small organelles, which are bathed in cell sap. The grana are composed of numerous microscopic leaflets with a large surface area exposed to the cell sap. They are packed with pigments of many colours, ('the autumn tints') as well as the green chlorophylls, and by this means absorb most of the incident light. The light rays set up a state of 'resonance', and by the system of 'resonance transfer' their energy too is transferred to chlorophyll a in a form which can be utilised in photosynthesis. Thus all these pigments, chlorophyll b included, act as 'antennae' for the chlorophyll a. The grana are also rich in the accessory enzymes and compounds necessary for electron transport and other reactions, and

Fig 16 ENERGY PATHWAYS IN PHOTOSYNTHESIS

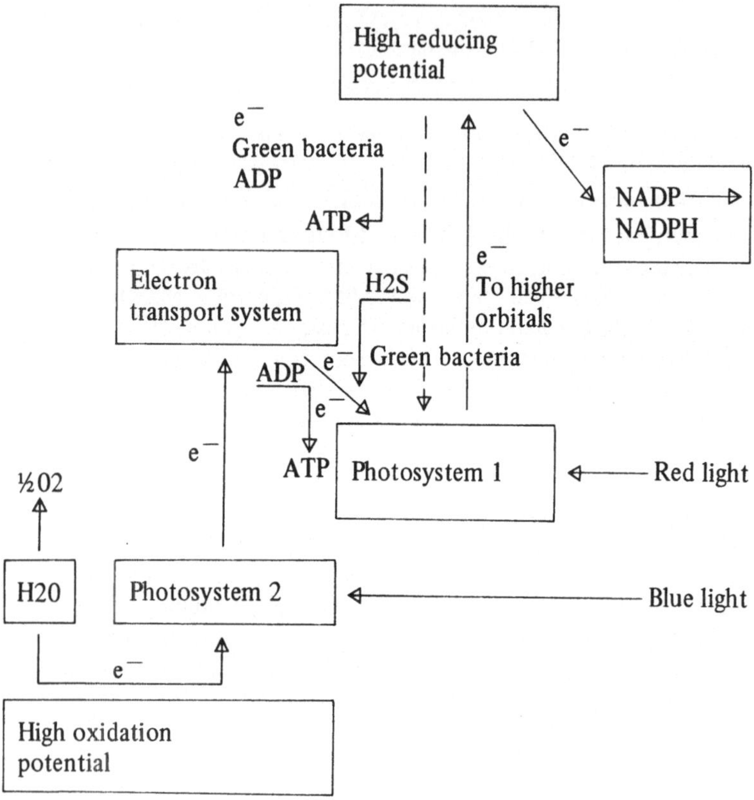

High reducing potential

e⁻
Green bacteria
ADP

ATP

NADP → NADPH

Electron transport system

H2S

e⁻
To higher orbitals

ADP

e⁻ Green bacteria

e⁻

ATP

Photosystem 1 ← Red light

½O2

H2O

Photosystem 2 ← Blue light

e⁻

High oxidation potential

Fig 17 THE GRANA ARE COMPOSED OF MANY FINE LEAVES WITH A LARGE SURFACE IN CONTACT WITH THE CELL SAP.
Fine structure of a chloroplast. The organelle is enclosed by a single membrane and contains numerous clusters of parallel membranes called grana which support the photosynthetic mechanism.

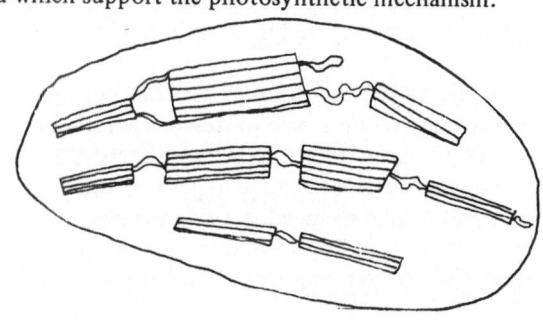

provide ready exchange of materials with the cell sap so that a rapid turnover is assured.

Nevertheless, the efficiency of light harvesting is limited by two factors, if not more. Photosynthesis is, of its nature, biphasic, requiring first the synthetic phase, and secondly a recovery phase while the synthesised products are exchanged for raw materials. Experimentally, its efficiency can be greatly improved in terms of rapidity of plant growth if the plant is exposed to rapidly intermittent light, allowing for optimum periods of synthesis and recovery. In the second place, photosynthesis is most efficient at certain optima of incident light and temperature. One might suppose that the high intensity light of desert areas would assist rapid photosynthesis. In fact, this is not the case, and many desert plants have silvery layers in their leaves which protect the chloroplasts from the full light intensity. Because of these two factors, theoretical calculations of the productivity of incident solar light on the earth's surface are of little value.

The end product of photosynthesis can be proteins or fats in addition to carbohydrates.[2] The process of photosynthesis is thus a matter of some complexity. In fact, the conversion of CO_2 into higher molecular compounds is not unique, since this can occur in the liver as part of normal metabolic processes powered by ATP. It is the use of the light photon as the energy source which makes the process unique. The whole of life as we know it today is dependent, therefore, on one green pigment. The chlorophylls themselves, however, belong to a well-known group of biological pigments, the porphyrins. The red pigments of the blood, responsible for the exchange of CO_2 and O_2 in the lungs and tissues, also belong to this group, but contain iron instead of magnesium. Primitive metazoan animals such as snails and tunicates experimented with other oxygen-transporting porphyrin pigments containing copper, niobium and vanadium.

From the high energy levels in carbohydrates and fats attained as a result of photosynthesis, the electrons when used in living systems drop back via ATP in a controlled and organised way stepwise to their lowest orbital positions; these positions occur in water, which is thus the excretory end product of energy in living systems. At each step energy is made available for life processes. Electronic energy could not, however, by itself be the basis of higher living systems, because it cannot conveniently be stored. It must be converted into chemical energy, and this is what happens when the energy is transferred to NADP and stored in the phosphate moiety.

Energy is, as we have seen, transferred at the end point of the photosynthetic process into carbohydrates and fat, which being insoluble can be accumulated. It is transferred back by way of ATP, and the reverse of photosynthesis occurs, 'oxidative phosphorylation'.

Electronic energy is the high biopotential of free electrons; chemical energy is the collective energy of the electrons forming the molecule. The single electron is mobile and can go to other atoms or molecules, that is to other orbitals of lower energy and with its negative electric charge will create an electric field. Chemical energy creates no such field and the electrons are immobile, being linked to molecules.

Energy Flows in Ecological Systems

The most characteristic features of life are these energy transformations, in which chemical energy is transformed into mechanical, electric, or osmotic work as in muscle, nerve, and glands respectively. How this is done is by no means simple and will not be discussed in detail here. Basically, the system depends on the balanced flow of electrons between systems of donors and acceptors mediated by enzymes. Donors possess loosely linked electrons; acceptors have spare capacity. The living cell is essentially an electrical device; in it the macromolecular structure is the framework in which the transduction of electrical energy into mechanical work takes place. The sources of electrons are the *activated* electrons of metabolism and are donated from the paired electrons of nitrogen, oxygen, and sulphur; the main acceptor group is found amongst carbon compounds.

The transmission of electrons from one substance to the other means the oxidation of one (the down path) and the reduction of the other (the up path). If solutions of reductant and oxidant are placed in beakers, the electrons can be made to pass from one to the other through a wire. The potential difference between the two solutions can be measured and gives information about the free energy exchange of the oxido-reduction which would take place if the solutions were mixed. The 'redox potential' allows us to calculate the energy changes taking place in the oxidation cycle. The redox does not measure the speed of a reaction, only the amount, and the energy change may differ with the medium. Mitochondria responsible for initiating and regulating these changes in cells, the medium is probably lipid, complex fatty compounds.

When energy needs to be recovered from the energy stores, the reverse of photosynthesis takes place, known as oxidative phosphorylation. In this oxygen is necessary, and the stored materials are again degraded to CO_2 and H_2O. For instance the group $- CH_2 -$ present in fats and other biochemicals plus $3O \rightarrow CO_2 + H_2O$; the CO_2, whether from plants or animals, is returned to the atmosphere. This is a similar process to that which occurs when coal or wood is burnt, but in combustion processes the stored energy is dissipated as heat, whereas in biological processes it must be diverted to processes of metabolism. In warm-blooded animals (homoiotherms), some

Fig 18 ATP could act as the 'transformer' of electric energy to chemical energy. The electrons would be trapped by the adenine and passed to the phosphate to be incorporated into the high energy phosphate bonds. (simplified formula)

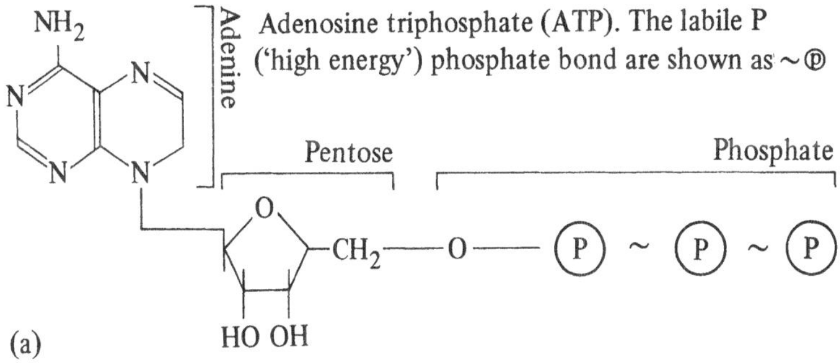

Adenosine triphosphate (ATP). The labile P ('high energy') phosphate bond are shown as ~ⓟ

(a)

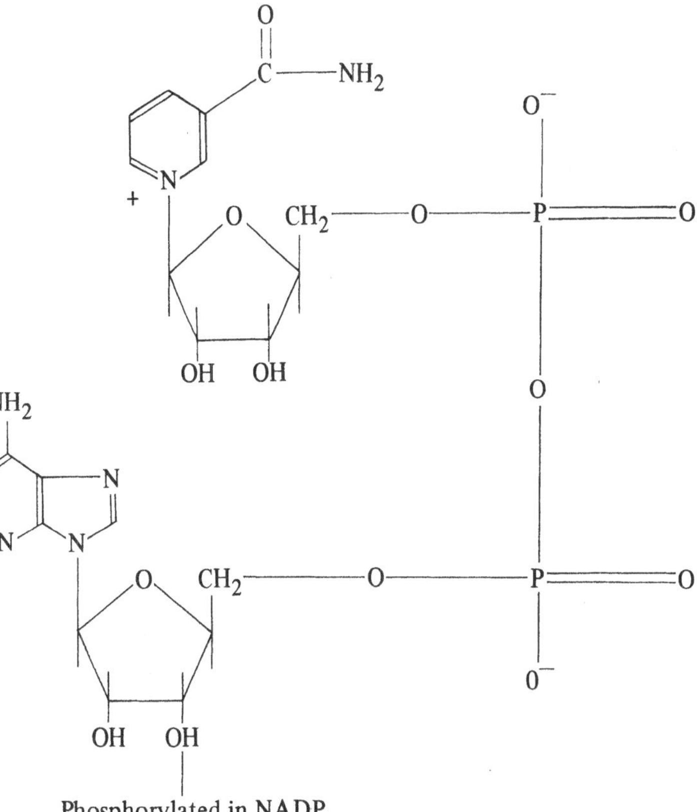

(b) Phosphorylated in NADP

energy is required for maintenance of body temperature and some of the heat produced is dissipated in maintaining this at constant level. This is why warm-blooded animals require a constant intake of food, whereas the larger cold-blooded animals (poikilotherms), such as crocodiles or pythons, feed only once a week or so and can survive for months on end without any food at all.

The use of energy for metabolic purposes depends on the extraordinary substance ATP, of which mention has already been made. ATP is a complex molecule, so arranged that it can act as a 'transformer' of chemical into electrical energy or vice versa. When fats or carbohydrates are broken down by enzyme action, the falling electrons, instead of creating heat, convert ADP (adenosine disphosphate) into ATP (adenosine triphosphate) in which the energy is trapped as chemical energy in the phosphate; it can then be delivered back as electrical energy, when and where it is required. ATP is unsuitable as the main energy store, but acts admirably as the medium of exchange. Biological 'currency' is stored in the high denomination notes, represented by fats and carbohydrates; ATP represents the small change of current transactions. ATP (adenosine triphosphate) consists of a compound base adenine, linked to a phosphate chain by a pentose (5 carbon sugar). This linkage is so arranged that various foldings are possible. Adenine is a good electron donor and the P atom, with unoccupied d-orbitals, would be a good acceptor. Thus electron exchange between the two could occur, if the molecule were folded in the right way. There is good evidence that the molecule is folded in such a way that the transfer could take place, and this explains the transformation of the electrical energy taken on by the adenine being converted to chemical energy in the phosphate.

It has been worth while to give this simple account of the trapping and use of radiant energy in biological systems, both because of its inherent interest as our basic life force and because it has been necessary throughout this work to make reference to it. For a more detailed account, the reader cannot do better than to study various contributions to the book *Biology of Nutrition,*[3] especially that by T.E. Cartwright.[4]

We have seen hitherto how light energy is used to convert simple chemical compounds, carbon dioxide and water, into storage elements, and how it can be recovered for building the structural elements of life, and made available for use in metabolic processes. Provided provision for decay is present in the form of lower saprophytic life forms, life could go on, and the elements could be recycled without higher life forms, apart from plant life. However, ecological systems are not constructed in this way, since there exist many higher life forms which use plant material as their energy source, and other life forms which live on the plant feeders. Mammals themselves are singularly ill-adapted to live on plant material. They were evolved in the Eocene and Oligocene epochs some 50 million years ago from primitive insect-eating ancestors. They do not have the digestive apparatus to deal with plant material, and lack the enyzmes necessary to split the higher polysaccharides such as cellulose, of which plants make their cell walls and largely consist. However, mammals have produced their herbivorous forms in the ruminants and the horse family, amongst the rodents, marsupials and even man. None of these could use plant material unless assisted by lower organisms which have not lost their power to attack plant material.

It is here of interest to see what happens to the energy of the photons trapped by the plants when introduced to a comparatively simple digestive system, such as that of a carnivore, which does not depend on lower life forms to do the job for it. The point is important in relation to the study of energy flows through ecological systems, which follows.

Before becoming available to build animal tissues, plant material must again be degraded by herbivores to comparatively simple compounds which can be absorbed through the bowel wall into the body. It is then resynthesised in the body, reconverted into storage and structural materials, and the energy is again released for metabolic activities. The digestive system of the flesh-eating animal again splits the proteins into their component amino acids; twelve carbon sugars, such as sucrose (cane sugar), are broken down to six carbon sugars; and fats are broken down to fatty acids and glycerol. Once absorbed, these materials again enter the metabolic pathways to be degraded, with release of their energy, to the metabolic end products of carbon dioxide and water.

Both the digestive and the anabolic (rebuilding) processes require considerable inputs of energy, so that at each stage a great deal of energy is lost. Furthermore, the whole process is subject to the First and Second Laws of Thermodynamics. A given quantity of organic matter contributes precisely the same amount of energy, whatever are the means used to degrade it; and when put to work an identical amount of heat will be produced whatever the nature of the work. No work can be performed without the conversion of some energy from a relatively non-random form (chemical or potential energy) to relatively random energy (heat) which is lost. We are, thus, going to find energy loss at each step in our ecological food chain, plant to herbivore, herbivore to carnivore, carnivore to scavenger, and from all these to organisms of decay. A further considerable loss of energy occurs also in animal excreta, since digestive processes never extract all the energy-containing materials available. This is of lesser consequence when excreta are

Fig 19 BIOMASS AND SOLAR ENERGY

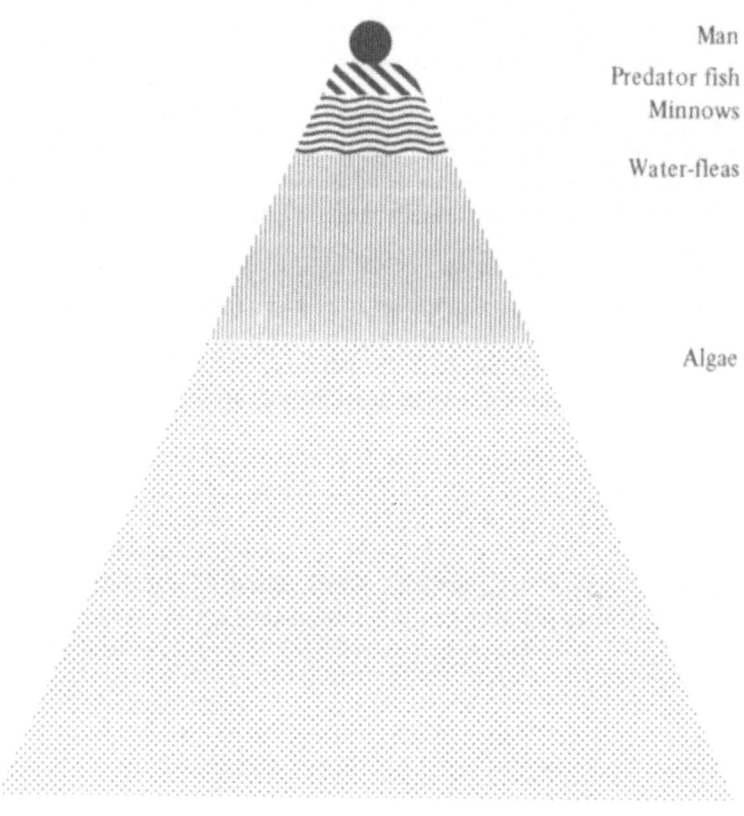

Man
Predator fish
Minnows

Water-fleas

Algae

returned to the land, but becomes more important when they are swept out to sea in the main drain. Obviously, when population pressures on a habitat are high, the most efficient management of the habitat becomes important.

Just how much energy is lost in the process is seen from the accompanying diagram (Fig. 19) from Hicks.[5] This shows at least that theoretical possibilities exist of more efficient use of energy resources, but at each stage in the food chain the more valuable elements become more concentrated, so that the supposed loss of efficiency is in part at any rate illusory, a point that is often overlooked. Animal products are expensive and the poorer human communities have mostly adopted diets preponderantly vegetarian; nevertheless they are a more valuable food source.

A natural ecosystem may be more or less productive, in the sense of the total biomass it will produce. Its total production will depend on the efficiency of its individual members. Basically, however, this will be a function of the climate and the soil, and its productivity can be improved by improving one or both of these, by supplying water or protecting surfaces from evaporation, or by providing missing nutrients to the soil. It could also be improved by replacing inefficient producers in the ecosystem with more efficient ones. For instance, there is a tendency today to suppose that the introduction of cattle, as the main source of animal protein, to tropical Africa was a mistake made by early immigrant peoples. It is argued that other indigenous herbivores, such as eland and buffalo, are natural and more economical exploiters of the ground cover and would cause less damage to it. Schemes have been suggested for replacing cattle with these animals, though I do not know of any studies made on the energy flows that occur when this is done. Nevertheless, it is certain that man has it in his power, by making proper studies of habitats and ecosystems, to improve energy flows and thereby gross total production. Agriculture could, therefore, lead to greater efficiency rather than less. The same remarks apply with equal or greater force to aquatic environments, especially estuaries. This has been shown, particularly by the Japanese in their fish culture ventures.

However, it is worthwhile looking at a few figures, drawn from Lamont C. Cole.[6] The supply of solar energy to the earth amounts to 13×10^{23} gramcalories per year, equivalent to a continuous power supply of 2.5 billion billion (American billion) horsepower. One third of this energy is immediately reflected back to space, chiefly by clouds. The rest is on loan to earth, until it is radiated back again. It serves to melt ice, to warm the land and oceans, to evaporate water, to generate winds and waves and currents. Only about four-hundredths of one per cent of the solar energy is used for photosynthesis. Of this, the plants use one-sixth for their own metabolism, leaving five-sixths available for animals and other consumers. Even of this, some 5 per cent is dissipated by forest and grass fires and burning of plant products for fuels. Of the plant material, herbivores succeed in converting only 50 per cent of the calories, and of this only 20 to 30 per cent is built into the protoplasm giving an efficiency of 10 to 15 per cent. The conversion efficiency of carnivores is around 20 per cent, because there are fewer indigestible residues.

Cole gives an interesting illustration taken from Lake Cayuga. Of every 1,000 calories stored by the lake algae, only 150 calories can be converted into protoplasm by small aquatic animals. Smelt, which feed on these animals, produce only 30 calories. If the smelt is eaten by a trout, the figure is reduced to around 6 calories. When the trout is eaten by a man, only 1.2 calories are left. Cole goes on to estimate that, if the

55

existing population of man were to feed exclusively on vegetable material, he would consume almost exactly 1 per cent of the total food resources of the earth. Since there are more than one million species of animals, he feels that this is an unreasonably high proportion. If man continues to demand a meat diet, this, of course, represents a much higher proportion of available resources.

One would like to see an analysis of energy flows through a highly industrialised society, such as that of the British Isles. In such a society, special features operate which may in the long run lack stability. Such are the importation of food and fertiliser from overseas countries. In addition, much residual energy and other valuable resources are wantonly swept to the sea in the sewage system. The latter is discussed by J.C. Wylie in his admirable little book *Fertility from Town Wastes.*[7]

Wylie, the Divisional Engineer of a Scottish burgh, was the architect of a sewage composting system. He succeeded in persuading his authority to give a trial to a disposal plant in which the garbage was composted with sewage sludge. The compost was sterilised by fermentation heat and was then sold for use on farms at good prices. Further extraction retrieved valuable minerals, chiefly phosphate and nitrogen, which are responsible for estuary pollution.

By recovery of waste materials in this way, industrialised countries could do much to redress an adverse balance in their energy accounts, and indeed in their balances of valuable minerals. They could also, with profit to themselves, avoid polluting their inland and inshore waters. While Wylie's initiative has been followed by some enlightened local authorities, argument still continues amongst local councillors as to whether a penny rate (the system in which the UK funds local schemes) should be spent on the necessary plant, a penny rate recoverable by the sale of the compost, or whether to spend a twopenny rate on shipping the garbage in barges to destroy the marine ecosphere.

I have touched briefly in this rather discursive chapter on natural minerals required in ecosystems. Further discussion of these and of the natural cycles by which they are reused will be left to a later chapter. Our main concern here has been to see how the sun's energy is trapped to become the driving force of life and how it flows out not only through living systems themselves but also through ecological systems. If man arrogates to himself the right to change the earth's ecosystems, he should at least acquaint himself with the best ways of controlling this energy flow for amelioration of his habitat rather than to its detriment.

NOTES

1 I.N. Healey, 'The Habitat, the Community, and the Niche', in R.N.T -W - Fiennes, *Biology of Nutrition,* Pergamon Press, Oxford, 1972

2 H. Davson, *A Textbook of General Physiology,* Churchill, London, 1959, Ch. XXII, 'Photosynthesis'
3 Fiennes, op.cit.
4 T.E. Cartwright, 'The Ways of Acquiring and Utilizing Energy', in Fiennes, op.cit.
5 Sir C. Stanton Hicks, 'The Nutritional Requirements of Living Things', in Fiennes, op.cit.
6 Lamont C. Cole, 'The Ecosphere', in R. Ehrlich, J.P. Holden, and R.W. Holm, *Man and the Ecosphere,* W.H. Freeman & Co., San Francisco, 1971
7 J.C. Wylie, *Fertility from Town Wastes,* Faber, London, 1955

Further Reading

R. Ehrlich, J.P. Holden, and R.W. Holm (eds), *Man and the Ecosphere,* W.H. Freeman & Co., San Francisco, 1971
E. Le Cren, and M.W. Holdgate (eds), *The Exploitation of Natural Animal Populations,* British Ecological Society Symposium, Blackwell, Oxford, 1962
A. Szent-Gyorgyi, *Bioenergetics,* Academic Press, N.Y., 1957
A. Szent-Gyorgyi, *Introduction to Submolecular Biology,* Academic Press, N.Y., 1960
A. Szent-Gyorgyi, *Bioelectronics,* Academic Press, N.Y., 1968

5 THE OCEAN/ATMOSPHERE SYSTEM AND 'CYCLING' OF THE ELEMENTS

Ecosystems may be wasteful and yet the energy flow through them enables increasing order to be produced from chaos. No less important is the flow of nutrients through these systems; they are the raw materials of which both individuals and the systems as a whole are constructed, and comprise the fuel the systems use. The ocean/atmosphere system, another flow system, recycles water resources. All life is based on water and cannot exist, except in a resting phase, in its absence.

As with life itself, the ocean/atmosphere system[1] requires a power source and, as with life, the power source is the sun. Whereas living systems use the visible rays of the sun, the ocean/atmosphere system uses the heat rays. Basically, water is evaporated from water masses, chiefly the ocean; the water vapour passes into the atmosphere; it is cooled in the upper atmosphere, forms clouds, and is reprecipitated as rain. When rain falls on land surfaces, some of the water finds its way directly into rivers and is returned to the sea. Some of the water is stored in lakes and reservoirs, in the polar ice, in the water table, or in the soil. The stored water may remain in the ground for long periods or be returned to the atmosphere by evaporation or transpiration from plants. On the face of it, this is a simple enough story, but the science of meteorology is by no means simple because air and water movements are influenced by so many unpredictable and unknown factors. In no two years are weather patterns the same, and even in waterless deserts occasional rainstorms occur. On a short-term basis, there are said to be eleven-year climatic cycles related to sunspot activity. On the medium term, there appear to be warmer and colder trends to the order of 400 to 500 years. On the long term, there appears to by no rhyme or rhythm which orders the appearance and disappearance of ice ages.

Since I am not a meteorologist, I shall not attempt to explain that which is obscure even to the experts. However, we may briefly look at the forces at work and review past and present climatic trends. Important forces are: (1) the heat of the sun and its position during the different seasons; (2) the spin of the earth which determines a predominantly westerly system of winds; (3) the gravitational effect of the moon on the earth causing tides; (4) the nature and direction of ocean currents; (5) the effect of land topography on air movements.

The sun's heat, being most intense at the ecliptic, causes moisture-laden warm air to rise; when it is cooled in the upper atmosphere, precipitation occurs. Thus a rain belt follows some weeks behind the sun's position. This belt is known as the inter-tropical convergence zone, and air movements occur from all directions into the zone which is thus cyclonic and constitutes the area of uncertain winds known by sailors as the doldrums. During the summer, much of the air which has passed to the upper atmosphere is transported to the polar regions, whence it is returned in very cold condition in the latter part of the winter as blizzards. At the same time, the westerly prevailing winds distribute the ocean moisture to the land masses. Over the great oceans, the maximum effect of the sun in causing an uprush of moisture-laden air may be so great as to cause hurricanes or typhoons.

Over uniform surfaces, air movements could be reasonably simple, but neither marine nor terrestrial surfaces are uniform. The oceans consist of great masses of water, moving this way or that in warm or cold currents, layered one above the other, downwelling and upwelling, diverted hither and thither by land masses and the shape of the ocean floor. To a great extent they are caused by the melting of Arctic and Antarctic ice by the sun's rays in the northern and southern summers. A cold current releases less water vapour into the atmosphere, and so is conducive to dry conditions. Cold currents of Antarctic origin pass along the west coasts of Africa and South America and result in desert or semi-desert conditions there. Conversely, warm currents cause moist conditions and high precipitation occurs over lands in their vicinity. This accounts for the moist conditions in countries such as Britain and Ireland whose shores are washed by the Gulf Stream.

Any factor which causes moisture-laden air to rise and be cooled, will also give rise to precipitation. Such factors are hills and mountains, and greater precipitation occurs, therefore, in hilly places. This effect is enhanced by the moist atmosphere in such places, the result of transpiration from the generous vegetational cover arising from the high rainfall. The vegetation also prevents excessive runoff and ensures absorption of the water into the soil, from which it seeps into the water table and feeds springs and wells. One of the surest ways of destroying a habitat is to cut down watershed forests, an activity to which man is much addicted. Precipitation is reduced; topsoil is removed by water erosion; and lower-lying land is destroyed by periodical severe flooding. At the same time, rivers come to flow only seasonally, and the water table becomes lowered.

Another factor is often adduced as a cause of climatic change, namely the composition of the atmosphere in respect of carbon dioxide. This problem is discussed at length by G.H. Plass.[2] An increase in atmospheric carbon dioxide is said to have a greenhouse effect tending to make the atmosphere warm and steamy, because the warm air is prevented from rising. It has been suggested that the burning of fossil fuels since 1860 has raised the atmospheric carbon dioxide from

0.028 per cent to 0.033 per cent with a consequent rise of the average temperature in the northern hemisphere by 1 to 2°F. If this were so, one might expect the opposite effect to have occurred in the Carboniferous, when the coal measures were laid down. This does not appear to have happened, and any effect such as this must surely be rather marginal. However, one factor of the ocean/atmosphere system that must not be overlooked is that of the seas in absorbing and regulating the levels of atmospheric carbon dioxide so as to maintain a stable equilibrium; reference has already been made to this (p.21).

Glaciation

The causes of the great glacial epochs, with their interstadials, are unknown. They are usually attributed to fluctuating intensity of solar radiation. There is no proof of this, but it is perhaps the most reasonable explanation. However, the effects of glaciation on land topography are far reaching. At the height of the last glaciation, the southern limits of the ice reached the Thames and to the south of the Alps. Below the permanent ice, the vegetation would be that of Arctic tundra. During the summer, the ice would melt and expose enormous areas of low Alpine-type vegetation, very well watered by streams from the melting ice and snow. The vegetation supported enormous herds of grazing animals, wild horses, mammoths, reindeer and so on, which moved south in winter and north in summer. Man himself was very much at home in this habitat and lived by hunting these animals, at which he was very successful.

The glaciation was accompanied in tropical areas by periods of heavy rainfall, known as pluvials. The equatorial forests extended far to the north of their present limits, and the Sahara was a well-watered plain, over which animals grazed and man also hunted. The ice only began to recede some 20,000 years ago, after which forests grew in man's former hunting areas, the wild life changed in character, and conditions became much less to the liking of our ancestors. It was this which forced them to take to agriculture and stock husbandry, and started the great social revolution which has been going on ever since.

During the glaciations, there was a drastic change in the ocean/atmosphere cycle. Myriads of tons of water were locked up in the polar ice, to an extent that much land was exposed where now there is sea. The English Channel was non-existent, England being joined to France, and much of the North Sea was dry land. Alaska was joined to Siberia, not only by ice, but by a land bridge also permitting the passage of wild life and of man (Eskimos and Red Indians) between the two continents. More than this, the weight of ice on some land surfaces, as in the Baltic, was so great as to tilt it downwards. Consequently, when the ice receded, though many land surfaces became flooded,

Table 2 *Chronology of the Glaciations During the Pleistocene Epoch*

Stages of glaciation	*(Years ago)*
III	25,000
Last glaciation II	72,000
I	115,000
Last interglacial	
Penultimate glaciation II	187,000
I	230,000
Penultimate interglacial	
Ante-penultimate glaciation II	435,000
I	476,000
Ante-penultimate interglacial	
Early glaciation II	550,000
I	590,000
Villafranchian, when earliest worked stone tools are found	1,000,000

This table shows glacial chronology in west and central Europe and covers the million years of man. During the period of the Pleistocene, increasingly sophisticated stone tools were made.

others relieved of the weight of ice actually rose and ceased to be submerged.

The earliest traceable glacial episode in earth history[3] occurred in late Pre-Cambrian times, 600–700 million years ago. In the succeeding Palaeozoic era, there was a temperate climate everywhere, succeeded in later Palaeozoic times some 250 million years ago by glacial conditions particularly in the southern hemisphere. The last ice age, which has affected us in modern times so much, occurred during the Pleistocene period, during which modern man evolved. The Pleistocene glaciation consisted of four glacial epochs separated by three interstadials. The ice first encroached during the early glaciation some 590,000 years ago, but had receded 40,000 years later. The second glaciation occurred some 500,000 years ago, and lasted 25,000 years. The third glaciation started some 230,000 years ago and may have lasted 50,000 years. The last glaciation started some 115,000 years ago and the ice was

61

retreating between 25,000 and 10,000 years ago. Typical arctic tundra still persists in Greenland and other northern areas. The changes that have occurred in the vegetation during the retreat of the ice have been determined with great exactitude by studies, particularly in Scandinavian countries. Not only has it been possible by geological studies to determine the limits of the ice and the glaciers and to assign dates to them, but by pollen studies it has been possible to ascertain the dates at which new vegetational successions followed the tundra. Much of the Sahara was still fertile and supporting nomadic herdsmen in early Egyptian times, and in parts plainly man's own efforts and those of his goats hastened the process of desiccation.

The end of the last glacial epoch is thus recent history. The floods that accompanied it are recorded in man's written history, and the myths and legends and taboos of an earlier period survive in cave paintings and in recollected superstitions and customs. Although the recession of the ice to its present limits seems to have occupied some 10,000 years, yet the climatic change which initiated it seems to have been quite sudden, and quite unconnected with any man-made activities, pollution of the atmosphere, and so on. The onset and retreat of glacial conditions are plainly due to natural phenomena, and one must suppose that the forces involved are far greater than any that man can command. A return of the ice would inevitably and inexorably drive man from his northern habitats, which he would be powerless to save. The point is stressed, because it is so seldom realised how altogether puny are the forces at man's command, even the most sophisticated, in relation to natural phenomena. No doubt man can destroy himself and temporarily pollute sections of the earth's environment; he can make little lasting impact.

Continental Drift

A true understanding of earth's past climates is, in fact, difficult to achieve. We see in tables compiled in accordance with the geological evidence that at times the whole earth has been warm and temperate even at the poles. Oil is found in Alaska and coal in Spitzbergen. How can this be, when the sun never shines during the Arctic and Antarctic winters? Surely these areas must always have been bitterly cold? Part of the explanation, though not the whole of it, is continental drift. That is to say that land masses were not always situated where they are today.

In brief, the land masses of the earth are not solidly anchored, but float on a layer of the earth's core which is known as magma and is in fact molten rock. We have seen that under great weights of ice, the land surfaces can be tilted and recover when the ice melts. It follows that continents can drift and in the course of time can occupy different positions in relation to each other and in relation to the poles and Equator. Originally, eastern America and western Europe and Africa were linked by their continental shelves, and Antarctica, Australia and India adjoined South Africa. There then existed one large land mass, Pangaea, with shallow seas scattered over it. This land mass was split by rifting, separating the Americas from Africa and Europe, but leaving the great land mass of Australia, Antarctica, South Africa and India, which is called Gondwanaland. This drifted towards the Equator, but then also split, India joining the Asian land mass and Antarctica moving to the South Pole.

The evidence for such drifting is too strong to be controverted, and the previous fusion of the continents is further attested by the distribution of animal life, for instance the presence of marsupials, such as opossums, in South America and Australasia but nowhere else. Thus the Antarctic land mass must at some time have been much nearer the Equator and, if we could move the $2 - 3$ mile depth of ice which now covers it, we should find fossil evidence of primitive mammal life, particularly of marsupials. Land masses, therefore, have not always been where they are now, and coal remains at the North Pole would not necessarily indicate a tropical climate in polar regions. Evidence of the great tectonic movements which accompanied the separation of the continents is still to be seen in our great rift valleys and mountain ranges, such as the Great Rift Valley in East Africa, the Himalayas, the Alps and the Andes. Much of this rifting still continued into the times when primitive man was making stone tools.

I may have seemed to have strayed from my initial subject, the ocean/atmosphere system. In fact, this is not so. The great ice ages, the drifting of land masses, and the position of the continents must inevitably affect earth's climates and the exchange of water vapour and other gases between the oceans and the land surfaces. Today, these upheavals are largely over and the earth's surface seems stable and placid. Nevertheless, earth's crust contains well-known areas of instability, such as the well-known San Andreas fault through California. Who knows when some other great upheaval might not occur along lines of weakness, with consequent climatic changes from alterations of ocean/land relationships? Catastrophic rains or a new ice age could well upset the nice balances which decree our pleasant climate.

Ecological Cycles

The flow of nutrients through living systems is also usually studied by means of 'cycles', and such cycles can be compiled for virtually all the elements that are of importance in the construction of living matter and for its metabolic processes. The most important of these are nitrogen, carbon, sulphur and phosphorus. However, before passing to the cycles,

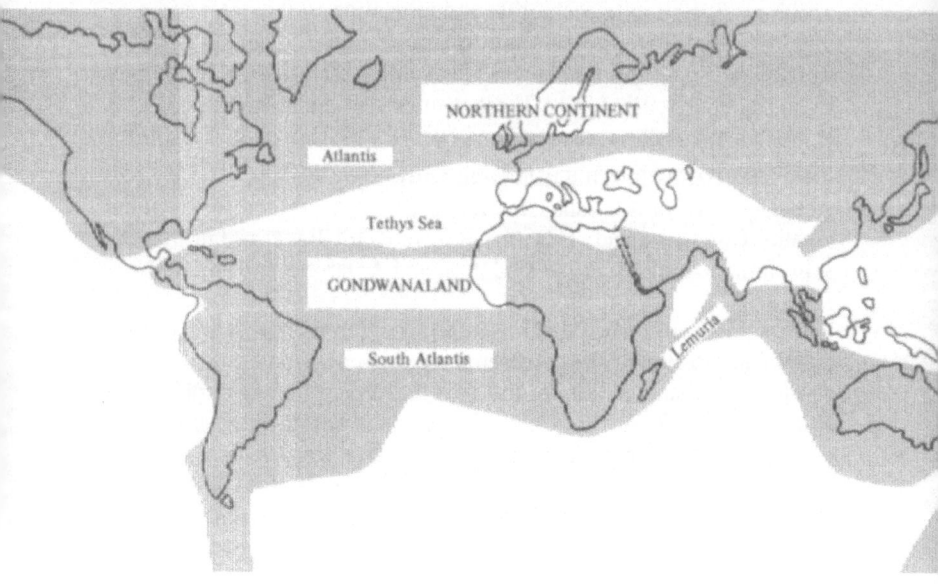

from Cloudsley Thompson *Terrestrial Environments*, p.4: Croom Helm, London, 1975

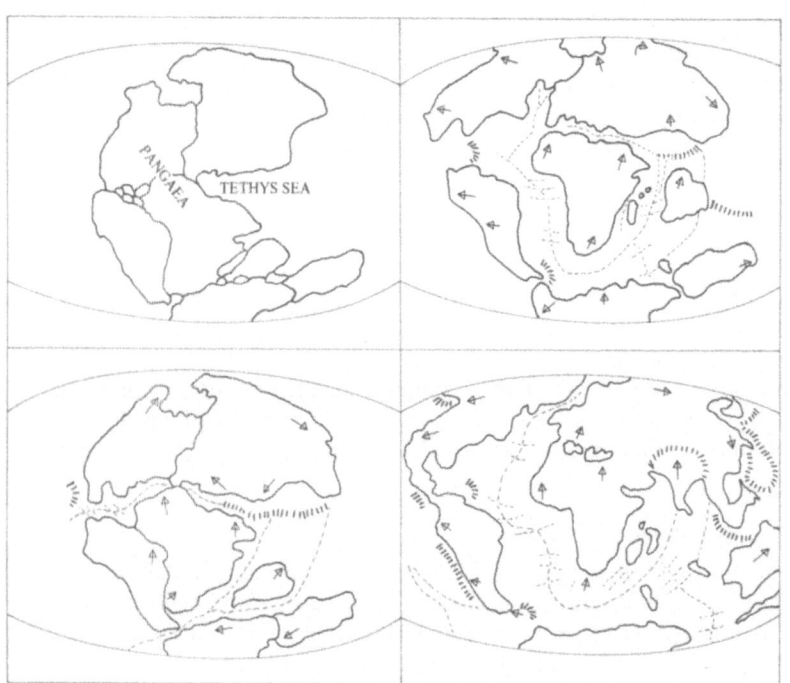

let us examine how these elements fit into living systems.

As a result of photosynthesis, carbon dioxide (CO_2) and water (H_2O) are combined to form carbohydrates. One such is the six carbon sugar glucose, of which the empirical formula is $C_6 H_{12} O_6$. The structure is simple:

$$
\begin{array}{cccccc}
OH & OH & OH & OH & OH & OH \\
| & | & | & | & | & | \\
C- & C- & C- & C- & C- & C \\
| & | & | & | & | & | \\
H & H & H & H & H & H
\end{array}
$$

There could plainly be variations on this theme; one such is galactose, the sugar of milk; another is fructose or fruit sugar. These are known as monosaccharides. When two monosaccharides are linked, a disaccharide is formed; thus glucose linked to fructose makes sucrose or cane sugar, the sugar we use in our tea or coffee. When further combined, sugars form polysaccharides; such are starch and the woody tissues of plants.

The second important range of biochemicals is that of the fats. These are built from fatty acids, of which the empirical formula is $CH_3 CH_2 CH_2 - COOH$. You can insert as many CH_2s as you like making a whole range of fatty acids with different names; amongst the more simple are acetic (the acid of vinegar), and butyric (the acid of butter). Again the arrangement is simple:

$$
\begin{array}{ccccc}
& H & H & H & H \\
& | & | & | & | \\
H - & C - & C - & C - & C - COOH \\
& | & | & | & | \\
& H & H & H & H
\end{array}
$$

When esterified (linked to alcohols), they form fats. In biological systems, they are linked to the alcohol glycerol to form vegetable or animal fats. In this form, fats are high sources of energy, and are the depot fats of animals. However, fats also occur in the unsaturated forms, in which they are of the greatest importance in body structure being known as essential fatty acids. In this form, some of the hydrogen atoms are missing, so that the compound is chemically unsatisfied and, therefore, capable of more complex linkages. The carbons in this case are linked by double bands where the hydrogens are missing; there may be only one double band or a number thus:

$$CH_3 - CH_2 - CH = CH - CH_2 - CH_2 - CH_2 - COOH$$

65

These unsaturated fats make the liquid vegetable oils such as olive oil or corn oil. They enter into the composition of all cells and in complex chemical form linked with phosphorus and other compounds form the white sheaths of nerve tissue. If not present in the diet in adequate amounts and the correct formulation, all sorts of serious results follow.

An amino acid is formed by inclusion of nitrogen in a fatty acid formula thus: $CH_3 . CH_2NH_2 . COOH$, to take the simplest. These acids are amphoteric, that is they can act as acids or alkalis, the carboxy group (COOH) being acid and the amino group (NH) being alkaline; they thus have a wide range of possible groupings. When combined, they form polypeptides; polypeptides combine to form proteoses; and proteoses combine to form proteins. There are twenty amino acids which occur in living systems, of which ten are dietary essentials. The proteins are the essential building components of the body — muscle, connective tissue, skin, horn and nails.

Apart from their importance as structural elements, many of the proteins play a part in metabolic activities as enzymes, without which the biological machine could not function. Amongst the most important are those proteins constructed of amino acids which contain sulphur in the form of a sulphydryl (SH) group. The sulphydryl group acts as a ready electron acceptor or donor, and is therefore necessary for energy pathways. Sulphur is, therefore, essential to living systems, and, being a somewhat rare element, its return to the pool by the sulphur cycle is of great importance.

Phosphorus also is an essential component of the energy system, having a number of functions. It is of particular importance, as we have already seen, because it is a component of the great energy transducer adenosine triphosphate (ATP). It is also an essential component of bones, nerves and other tissues.

A great many other elements are, of course, required both as structural elements in a living body and in metabolic processes. All of them must be present in the diet or mineral additives, or in salt or salt licks. However, the importance of ecological cycles resides to the greatest extent in the four mentioned, so we shall study them in these, to show the principle of what happens.

Oxygen Cycle

Let us first look at oxygen, which we have already encountered as a pollutant of the atmosphere in early times, when life first appeared on earth. Even today, oxygen is not a *natural* constituent of the atmosphere. By this, I mean that, if earth should again be bereft of its vegetation, oxygen would again disappear from the atmosphere. Expressed in another way, almost all the atmospheric oxygen is deposited there by plant life. Some 70 per cent of the oxygen is derived from the marine plankton, that is by the one-celled plant life of the oceans. The remaining 30 per cent is derived from plant life on land, chiefly forest trees; grass and other soil covers are less efficient.

It is a tribute to the photosynthetic efficiency of these plants that the level of atmospheric oxygen remains so remarkably stable. Nevertheless, fears have been expressed by a number of competent scientists that this will not always be so, for three main reasons. First, man's requirements of oxygen have increased to such an extent that oxygen levels in the atmosphere over major cities already show a deficiency of some 6 per cent, compared with the surrounding countryside. This is partly due to the oxygen required for respiration and partly to oxygen used in burning fuel for heating, transport, factories and other purposes. Secondly, the actual removal of plant cover in cities, roads, airstrips and other areas, together with forest clearance, has limited the power of plants to replace oxygen from land surfaces. Thirdly, any serious reduction of marine phytoplankton could lead to a reduction of photosynthesis of such an extent as to threaten oxygen reserves. In limited areas, such reductions of phytoplankton have occurred as a result of water pollution by detergents and chemicals, such as DDT and dieldrin.[4] Furthermore, closed waters as in the Great Lakes have been rendered virtually sterile because of the introduction of nitrogen in the form of its oxides, nitrites and nitrates and of ammonia from sewage and agricultural fertilisers washed into rivers from the fields. We shall look at this further when we study the nitrogen cycles. There is no immediate threat to the phytoplankton on a worldwide scale, but evidently the indiscriminate discharge into the sea of substances that could harm it requires to be controlled.

The actual oxygen cycle is a simple one. Plants in photosynthesis build carbohydrates from atmospheric carbon dioxide and water; during the process, excess oxygen is released into the atmosphere. The oxygen in the atmosphere is used in the respiratory processes of living things, including plants, and in the burning of fuel. The end point of these processes is carbon dioxide or carbon monoxide, which is converted in the atmosphere to carbon dioxide. The atmospheric carbon dioxide is reused by plants.

Carbon Dioxide Cycle

The carbon dioxide cycle, also, is very simple. In spite of this, as we have seen, the CO_2 content of the atmosphere remains extremely constant over periods of millions of years. It has been claimed, as already stated, that there has been a small rise in recent years as a result of the burning of fossil fuels, and that from a greenhouse effect the average temperature of the northern hemisphere has risen by a few degrees. It is also argued that this association may be erroneous, or alternatively that it is offset because of an increase of dust and other solid particles, which have the effect of insulating the earth's surface and thus cooling it. Is is thus undecided whether the extinction of life will come because the earth will be an inferno, or whether it will become entirely glaciated!

At any rate, carbon as one of the basic raw materials of life is not in short supply. It is plentiful in the atmosphere as carbon dioxide, and in the seas as dissolved carbon dioxide, as sodium carbonate and bicarbonate, and both in sea or land combined with calcium in the form of chalk, limestone, marble, gypsum and so on. There is also a great deal of carbon still locked up in fossil fuels, such as coal, oil, peat and lignite. It is the relationship of the seas to carbon that maintains atmospheric carbon at constant levels. When these levels tend to rise, more carbon dioxide becomes dissolved in sea water, and when the level in sea water rises carbonic acid is formed and this is combined with sodium ions to form carbonate and bicarbonate of soda. The seas are so vast that the effect on the composition of sea water is negligible. However, there appears to be some time lag in the process, and excess of carbon dioxide excreted into the atmosphere is not immediately removed; hence the small present-day increase of atmospheric carbon dioxide. It is relevant to remark that fossil fuels are expected to last at present-day levels of consumption for no more than a hundred years or so, and there appears to be no reason why the system cannot absorb them without serious consequences.

The cycle, as said, is inherently simple. Carbon dioxide is added to the atmosphere by the exhalations of animals (and of plants during the hours of darkness), by the combustion of fuels, and by the decomposition of dead organic material by microbial action. It is abstracted from the atmosphere by plants in the course of photosynthesis. Any imbalance in the process is corrected by the effect of sea water.

Nitrogen Cycle

Unlike carbon, nitrogen is a commodity in short supply, though equally with carbon it is an essential building material of living things. It would be more accurate to say that it is in short supply in a form that is available to living things. Higher plants obtain their nitrogen through their roots, not directly from the air, in which of course it is abundant. Moreover, it is only available for use in the form of nitrates, the oxidised form containing three oxygen atoms. Thus a number of complicated reactions occur in the soil which prepare nitrogen for plant use, the sources being: (1) atmospheric; (2) from broken-down organic debris (humus); (3) from artificial fertilisers.

Farmers using modern methods find it easier and more profitable to supply their crops with nitrogen by way of artificial fertilisers. It is not yet clear whether such systems, though in the short run profitable, can be sustained indefinitely. They include a number of undesirable features which may in the long run be self-defeating. The systems include the intensive use of the so-called agricultural chemicals for control of plant parasites and pests. The result is a virtual sterilisation of the soil, which is used as an artificial culture medium for holding the plant nutrients found to be necessary. We have already seen that a natural fertile soil is a living entity, crawling with life of different kinds and an ecosystem of its own. The living matter in soils processes organic residues, making them available for reuse; it lightens the soil and aerates it, so that atmospheric nitrogen is made available, and it provides the crumb structure, which permits water percolation and prevents erosion by wind and water.

It is well known also that top dressings of nitrogen and other substances are carried away by surface run-off into streams, lakes and reservoirs. Thirty per cent or more of fertilisers are quite normally lost in this way. The excess nitrogen has a most unfortunate effect in enclosed or slow-moving waters, which is known as 'eutrophication'. The end result is to remove the oxygen from the water, making it incapable of supporting fish and other creatures requiring oxygen. It happens in this way. Water algae are avid for nitrogen, and when this is supplied in excess their numbers multiply enormously to give 'a bloom'; by photosynthesis, they increase the oxygen in the water, and this results in the production of very large numbers of small organisms, which feed on them. These organisms quickly use up all the oxygen with the result that they and all other oxygen users die and the waters become a polluted, stinking mass of decaying material. Recovery from this condition may take many years. As a result of agricultural run-off together with sewage deposition, Lake Michigan is now a dead lake and may have gone past the point of no return.

In nature and still, fortunately, in our gardens, nitrogen problems are handled according to ecological laws evolved over many millions of years. Organic material suffers decay as a result of microbial action. Nitrogenous material is converted by one set of organisms to nitrites (oxide of nitrogen with 2 0 atoms), thence by another set of organisms to nitrates (with 3 0 atoms) available for use by plants. In well-aerated soils, atmospheric nitrogen is 'fixed' by other soil-living organisms, including those that are found in root nodules of conifers and leguminous plants. At the same time, organic debris forms humus and a rich tilth suitable for plant growth is produced. Some of the decayed material is converted to ammonia, thence to the oxides of nitrogen, both of which find their way back to the atmosphere to form nitrogen gas.

Nitrogen, as nitrate, is rather a rare material and very expensive to produce by synthetic means from atmospheric nitrogen. The tragedy of the present position is the wasteful and harmful way in which this rare material is being used. Not only does this apply to its use as fertiliser. Sewage systems are equally to blame. It is madness, and should be on the social conscience of industrial man, that the nitrogenous residues of urban centres should be swept out to sea whence they are irrecoverable instead of being processed for recovery of vital materials.

Sulphur Cycle

Another rare material essential to life is sulphur, and this is being equally wastefully exploited. The cycle is basically rather simple. Sulphur, as we have seen, is important in metabolic processes and present in all organic debris, from which it is released during decomposition as H_2S (hydrogen sulphide). Provided oxygen is present, this gas becomes oxidised to form sulphate. In this form it is again assimilable by plants and no problems arise. In the absence of oxygen, as may occur in marshes, estuaries, or thick reeds, the H_2S accumulates, giving rise to foetid conditions with a strong smell of rotten eggs. In such situations, it is eventually used metabolically by anaerobic sulphur-loving bacteria and photosynthetic green and purple bacteria. Sulphur in the form of sulphur dioxide is an important element of the smoke arising from coal burning. In this form it is dangerous to health, particularly in fogs or smogs, and a serious atmospheric pollutant. The discharge of sulphur, into rivers or estuaries as well as into the atmosphere, cannot be afforded. The world's sulphur resources are limited, and in the form of sulphuric acid it is of commercial importance. Its recovery from sewage or smoke presents no great difficulty, and the continuing folly of polluting water and atmosphere with this scarce element is incomprehensible.

Like sulphur, phosphorus is of great importance in metabolic processes and left to itself recycles by way of absorption by plants from organic debris in the soil. However, there do exist in the world limited deposits of phosphate in rocks and so on, and farmers find it more economical to treat their land with phosphate fertilisers than to rely on natural cycles. Excessive run-off from farm lands, together with phosphate in sewage and derived from phosphate detergents, assists the eutrophication of waters, besides squandering an important resource. Phosphate has been an important element in the destruction of the ecosystems of Lakes Michigan and Superior.

Cycles can be constructed for other elements important in metabolism, such as iron, copper, iodine, manganese, and so on, but

Fig 22 CARBON CYCLE

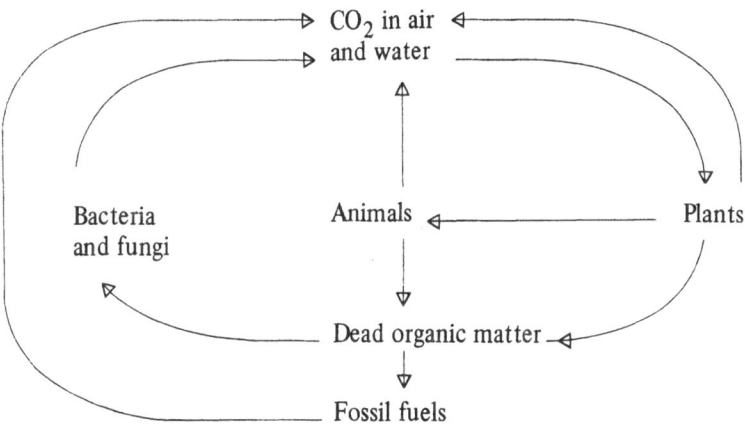

Fig 23 NITROGEN CYCLE

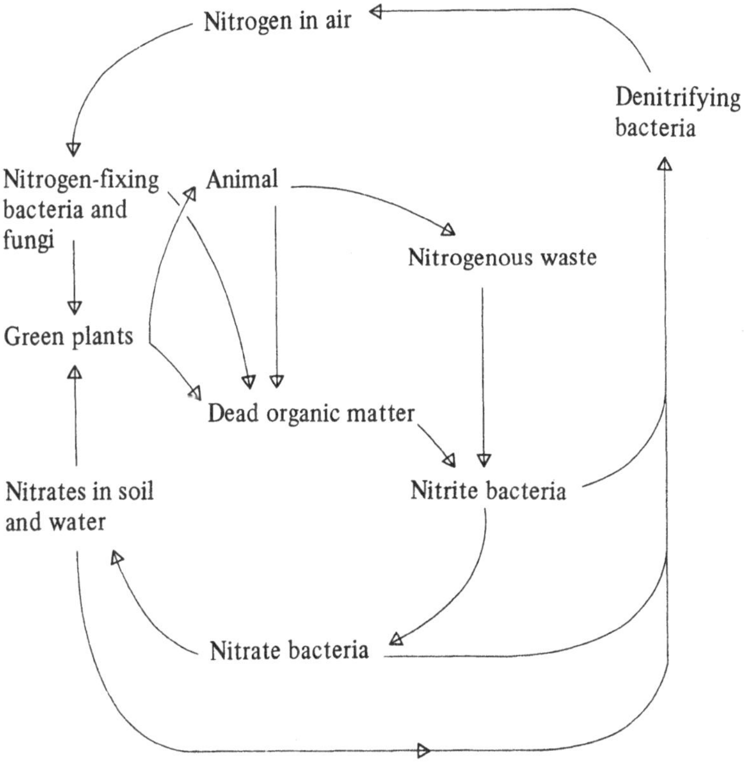

Fig 24 SULPHUR CYCLE

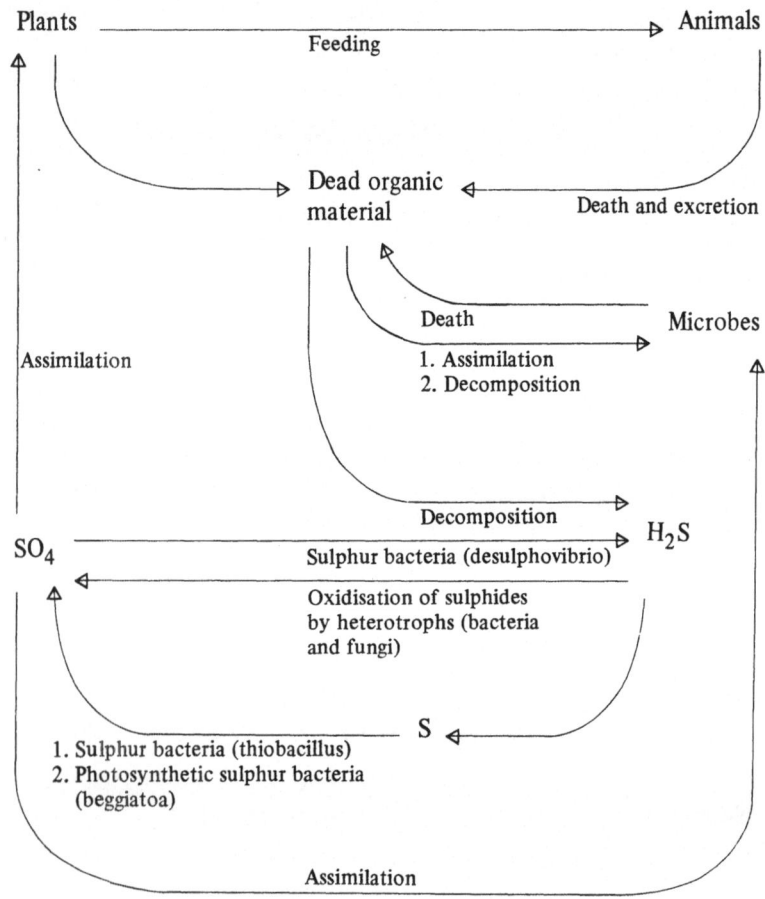

enough has been said to emphasise the importance of these cycles in nature's ecosystems and of the ways in which man is disrupting them. One of the most important problems of modern life is that of recovering some of these elements from waste materials and of returning them to the biological flow system. Mishandled, as now, these valuable materials become dangerous pollutants. Paradoxically, their recovery can be profitable.

We now pass in the next chapter to study the ways in which life forms interact with each other to form an ecological unit, the 'associes'.

NOTES

1 R.N.T -W- Fiennes, 'The Evolution and Colonisation of Habitats', in R.N.T -W- Fiennes, *Biology of Nutrition,* Pergamon Press, Oxford, 1972
2 G.H. Plass, 'Carbon Dioxide and Climate' in R. Ehrlich, J.P. Holden, and R.W. Holm, *Man and the Ecosphere,* W.H. Freeman & Co., San Francisco, 1971
3 E.C. Olsson, 'Climatic Change and its Influence on Life and Habitat', in Fiennes, op. cit.
4 G.M. Woodwell, 'Toxic Substances and Ecological Cycles', in Ehrlich *et al.,* op. cit.

6   THE INTERDEPENDENCE OF LIFE FORMS – THE 'ASSOCIES'

The ways of life and adaptation to terrestrial and aquatic environments of life forms are reviewed in subsequent volumes of this series.[1] Here it is necessary to survey the ways in which the members of a habitat interact with each other.

Modern works on ecology largely ignore the pioneer studies and theories put forward by the British zoologist Alfred Russell Wallace, who proposed the theory of natural selection at the same time as Charles Darwin gave his evidence of the mode of operation of evolution. Wallace's *Geographical Distribution of Animals,* published in 1876, was in fact the first major work on ecology, though that term had not then been invented. He pointed to the importance of natural and physico-geographical barriers in determining the distribution of plants and animals and to the results occurring when these barriers are removed. In particular, he drew attention to the effects of the long glacial epoch in dividing the world into faunal regions, each with a defined fauna and flora developed in isolation. However, the recession of the ice led to a redistribution of these regions and a tendency for a mixing of faunal types, with far-reaching ecological effects.

He puts so well some of the lessons this book attempts to expose, that it is worth our while to quote the eight principles he propounds:

(1) If the dry land of the globe had been from the first continuous, and nowhere divided up by such boundaries as lofty mountain ranges, wide deserts, or arms of the sea, it seems probable that none of the larger groups (as *orders, tribes,* or families) would have a limited range; but, as is to some extent the case in tropical America east of the Andes, every such group would be represented over the whole area, by countless minute modifications of form adapted to local conditions.

(2) One physical barrier would, however, even then exist; the hot equatorial zone would divide the faunas and floras of the colder regions of the northern and southern hemispheres from any chance of intermixture. This one barrier would be more effectual than it is now, since there would be no lofty mountain ranges to serve as a bridge for the partial interchange of northern and southern faunas.

(3) If such a condition of the earth as here supposed continued for very long periods, we may conceive that the action and reaction of the various organisms on each other, combined with the influence of very slowly changing physical conditions, would result in an

74

almost perfect organic balance, which would be manifested by a great stability in the average numbers, the local range, and the peculiar characteristics of every species.

(4) Under such a condition of things it is not improbable that the total number of clearly differentiated specific forms might be much greater than it is now, though the number of generic and family types might perhaps be less; for dominant species would have had ample time to spread into every locality where they could exist, and would then become everywhere modified into forms best suited to the permanent local conditions.

(5) Now let us consider what would be the probable effect of the introduction of a barrier, cutting off a portion of this homogeneous and well-balanced world. Suppose, for instance, that a subsidence took place, cutting off by a wide arm of the sea a large and tolerably varied island. The first and most obvious result would be that the individuals of a number of species would be divided into two portions, while others, the limits of whose range agreed approximately with the line of subsidence, would exist in unimpaired numbers on the new island or on the mainland. But the species whose numbers were diminished and whose original area was also absolutely diminished by the portion now under the sea, would not be able to hold their ground against rival forms whose numbers were intact. Some would probably diminish and rapidly die out; others which produced favourable varieties, might be so modified by natural selection as to maintain their existence under a different form; and such changes would take place in varying modes on the two sides of the new strait.

(6) But the progress of these changes would necessarily affect the other species in contact with them. New phases would be opened in the economy of nature which many would struggle to obtain; and modification would go on in ever-widening circles and very long periods of time might be required to bring the whole again into a state of equilibrium.

(7) A new set of factors would in the meantime have come into play. The sinking of land and the influx of a large body of water could hardly take place without producing important climatic changes. The temperature, the winds, the rains, might all be affected, and more or less changed in duration and amount. This would lead to a quite distinct movement in the organic world. Vegetation would certainly be considerably affected, and through this the insect tribes. We have seen how closely the life of the higher animals is often bound up with that of insects; and thus a set of changes might arise that would modify the numerical proportions, and even the forms and habits of a great number of

species, would completely exterminate some, and raise others from a subordinate to a dominant position. And all these changes would occur differently on opposite sides of the strait, since the insular climate could not fail to differ considerably from that of the continent.

(8) But the two sets of changes, as above indicated, produced by different modes of action of the same primary cause, would act and react on each other; and thus lead to such a far spreading disturbance of the organic equilibrium as ultimately perhaps to affect in one way or another, every form of life upon the earth.

He subsequently classified the faunal regions of the earth, as determined in the glacial epoch, as under:

Regions

Neogaea	(Neotropical	Austral zone	Notogaea
	(Nearctic	Boreal zone)	
	Palaearctic)		) Arctogaea
	Ethiopian)	Palaeotropical)	
Palaeogaea	Oriental)	zone)	
	Australian	Austral zone	Notogaea

The disruption of these zones since the end of the glacial epoch had depended first on climatic change; secondly, on the changed ecology of man, itself a result of climatic change. It is this that must now be studied, but first we must look at some facts of ecological stability, interference with which has led to the kind of instability that Wallace forecast. The main changes to be expected with climatic shift will be these:

1. Flora and fauna species may be no longer adapted to the new environment and new unoccupied ecological 'niches' may become available for colonisation. These vacancies will be filled either by adaptation – by natural selection – of the pre-existing life forms, or by invasion of the habitat by life forms from elsewhere. In practice, the latter has been predominant during the human era, and favoured by human activities. The results of invasion may be far-reaching and of long duration, as Wallace predicted.

2. New species invading the habitat may thrive to such an extent as to cause its deterioration. Instances are the introduction of coypu and mink into Britain, the grey squirrel into Britain, the rabbit into Australia, and goats into areas such as the Sahara fringes. The depredations of such animals may have very serious consequences leading to a complete change of the character of the habitat. For instance, goats – and cattle for that matter – will devour tree seedlings,

76

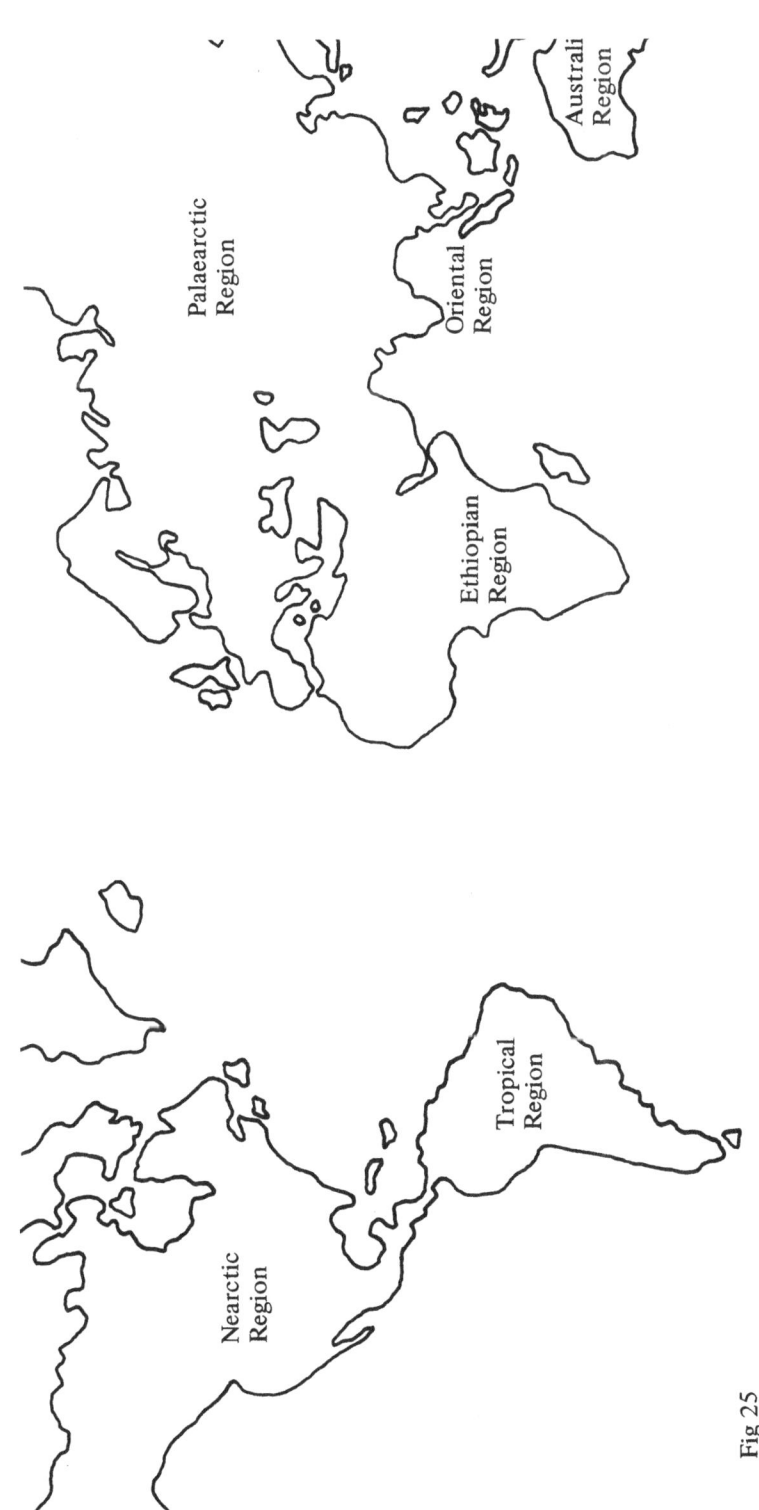

Fig 25

so that a forested habitat will become progressively grassland and – as has happened in areas of the Sahara – desert. Climatic changes can follow with desiccation, erosion, a fall in the water table and so on. The destruction of water-shed forests by man himself constitutes just such a challenge to the environment. The replacement of pre-existing species, of course, may lead to their extinction, and a great many animal species have indeed become extinct since the last glacial epoch.

3. The situation has been made worse by man's activities. This is especially the case where major predators such as wolves are exterminated. Their removal permits the multiplication of prey species to such an extent that natural forage becomes scarce or crops are destroyed. Man will then exterminate the culprits and the fauna of the land becomes the poorer. An example of this is seen in Britain, where a number of species of deer have been reintroduced. They have multiplied so fast as to be a nuisance and their numbers are controlled by periodical shooting of the excess population. A further instance of the results of man's interference is seen where, in order to preserve species threatened by the spread of human habitation, game parks are created. These parks favour the increase of static species but are inimical to those, such as elephants, which normally are seasonally migratory. So either the migratory species perform poorly in competition with the static, or alternatively, in order to sustain themselves on the restricted habitat, they virtually eat everything in sight and the habitat is destroyed. In this case again, man must intervene to control their numbers.

4. The invasion of a habitat, altered as a result of climatic change, is not made only by animals and birds. Insects, which carry disease, infiltrate also, and new parasites and disease germs of the invading species are also introduced. Even where an existing species is able to compete successfully, it may be weakened by the natural parasites of the invading species and succumb to the challenge. The disastrous decline of the red squirrel population in Britain in face of the grey squirrel is believed to have been as much due to its susceptibility to grey squirrel diseases as to direct competition.

5. Once the numbers of a threatened species fall below a certain limit, there occurs a point of no return from which revival is almost impossible. Above a certain lower limit of numbers, the population will thrive and in favourable circumstances numbers will recover. Below this point, the stock deteriorates, numbers diminish still more and the stock is likely to be extinguished. The reasons for this are not fully understood, but genetic defects incline to appear and to be perpetuated so that the constitution of the population is undermined. This occurs with human populations also, as is seen amongst such isolated communities as those of Tristan da Cunha. Even amongst such relatively

large populations as the whites of South Africa, unfavourable results from the perpetuation of genetic defects can be traced and they have been important in recent years in rural communities in European countries such as Britain, where the people were much inbred.

6. Ecological problems have also been intensified both in the human race itself and in the farm animals and crops raised, because of monoculture. Man's agriculture demands that crops be intensively cultivated, whether cotton, wheat, fruit, maize or anything else. This is directly opposed to the dictates of ecology, which demand diversity. In this way, while the desired crop is raised, the total potential biomass of a region is never realised. Moreover, parasites, pests and diseases are enabled to multiply in a geometrical progression and must be controlled if the crop is to be raised successfully. Hence the use of chemicals, bird scarers, and other devices to control these pests. It is not only with crops that these problems arise, because farm animals, particularly poultry (chickens and turkeys) intensively reared, are at especial risk to diseases and parasites, and these must be controlled by drugs and vaccines. Urbanised man himself is in the same situation, and epidemic diseases run quickly through crowded populations.

Once ecosystems are disturbed, whether by natural causes such as climatic change, or by invasion of life forms from outside, or by interference by man, far-reaching effects are the result, and these may be long lasting and deleterious. So long as an ecosystem, developed over thousands of years, is undisturbed, its numbers are self-adjusting. The herbivores are kept in check by the predators; the predators only kill for food not for the lust of killing; and the parasites and disease-producing organisms have become mutually adjusted with their hosts, so that serious epidemic diseases do not make inroads on the denizens of the habitat. The members of the ecosystem adjust their numbers by various devices, such as the well-known territorial system. In dry and cold seasons, those creatures which cannot find sustenance survive by a number of devices such as aestivation or hibernation, migration as with many bird species, or in the case of insects and annual plants in resting stages represented by fertile eggs or seeds.

Interdependence of Life Forms

Here we are concerned mainly with those rules of ecology which may be overthrown with undesirable results, and we need to study the mutual dependence of life forms on each other, that is commensalism, symbiosis, and their extension, namely disease.[2]

No forms of life can exist in isolation. This must be sufficiently clear from what has already been written. The stability of the ecosystems depends entirely on that of its members and their relationships with each other; on the organisms which promote decay, and thus ensure recycling of the elements, as much as on the primary and secondary producers. Throughout nature, there occur also innumerable instances of living things which combine together to make their living, and which could not flourish so successfully in any other way. This is known as 'symbiosis'. The most frequently quoted example is that of the lichens, seen everywhere on stone walls, etc., as pioneer life forms on surfaces which other plants are unable to colonise. By their activities, an arid and uninviting surface can be colonised and gradually eroded by the action of carbonic acid, so that crevices are formed to hold debris to become a sort of soil which secondary pioneers can invade.

Symbiosis is particularly prevalent in life forms on coral reefs. In such situations, corals, anemones, sea worms and even some protozoa contain chlorophyllous phytoplankton, known as zooxanthellae, in their actual bodies, and it has been suggested that in this situation they are responsible for much of the primary production on coral reefs.[3]

Now, mammalian herbivores in terrestrial situations are descended from insectivorous or carnivorous ancestors. Although they live on plant material, their digestive systems are not able to split cellulose and other higher polysaccharides, such as woody materials, lignin, and so on, which they consume. These functions are performed for them by bacteria and other micro-organisms in capacious fermentation compartments developed in their digestive systems. The most complex of these are the multiple stomachs of the ruminants, known as the rumen, the reticulum, the omasum and the abomasum. A similar development has occurred in the hippopotamus and the leaf-eating colobine monkeys. In horses, and other herbivorous animals, the fermentation chamber has been developed in a capacious large bowel, in particular the caecum. If the micro-organisms responsible are prevented from developing in the rumen of a calf, or are suppressed by antibiotic treatment, the animal cannot digest its food and will die. Some of these micro-organisms, furthermore, synthesise vitamins, particularly of the B group, and if eliminated by antibiotics vitamin deficiencies occur.

Apart from these beneficial organisms, all wild-living creatures harbour a variety of lesser creatures within their bodies, worms, protozoa, bacteria and possibly even viruses, which while doing them no harm do not positively contribute to their well-being. Such are known as 'commensals'. The commensals have a dual importance. First, in times of stress, illness, or malnutrition, they are inclined to become out of control and to cause symptoms of disease. Secondly, while harmless as a rule to the primary host, they may be dangerous disease-causing organisms to a secondary host, particularly one of a related but distinct species. In the first sense, their activities are probably beneficial to the tribe as a whole, because they tend to cause the deaths of weaker members not well attuned to them or to cause epidemic deaths when the group is over-populated, malnourished, or stressed, so helping to reduce numbers to a more favourable level. In the second sense, they only become important in circumstances of new introductions of animals to an ecosystem, or where an ecosystem is subject to change as discussed above.

A stable population, from tens of thousands of years of acclimatisation, is normally in balance with its parasites, and factors of disease are relatively unimportant. The heavy toll taken by disease both in human and animal populations in present days is the result of the disturbance of naturally developed ecosystems, and the introduction of strange parasites and diseases on the one hand, and of strange animals, including man, to new areas on the other. Monocultures, both human and other, favour the spread of these diseases. In addition, any activity which alters the environment may permit parasites, formerly under natural control, to gain ascendancy. The spread of Dutch elm disease is attributed to the destruction of hedgerows to permit combine harvesting. The natural control of the beetle which causes the disease is by woodpeckers, which delve into the bark for them and eat them. The woodpeckers are threatened by the removal of their hedgerow habitat.

The dangers of unaccustomed diseases in related species is shown by many instances which could be quoted, as for instance between man and other primates.[4]   The best known of these is that of myxomatosis.[5] Myxomatosis is a naturally occurring disease in the American cotton-tail rabbits, in which it causes transitory skin tumours. When European rabbits become infected, it is virulent enough to cause the death of more than 98 per cent. When infected rabbits were introduced deliberately into wild rabbit populations in Australia and Britain, it almost led to their extermination, but not quite. As is well known, the rabbits are returning and the death rate from myxomatosis has diminished, although the disease still serves to limit their numbers. Many diseases, too, which are natural to monkeys are very dangerous indeed to human beings, and the greatest care has to be taken in handling them. The well-known disease caused by B. virus, one of the herpes group of viruses, is a case in point. While infection of humans is fortunately rather rare, only one patient so affected has survived and he only with his mental faculties permanently impaired. Some of the natural herpes viruses of monkeys are also very dangerous to other monkeys, and in some they cause a deadly form of cancer. This, too, is a point of great importance, because it is suspected that some human cancers are caused by viruses, including the leukaemias.

Table 3 *Summary of Geological Calendar with Approximate Ages in Millions of Years*

	ERA	PERIODS		AGE
PHANEROZOIC	CENOZOIC		Epochs	
		Quarternary	Recent	
			Pleistocene	
				2.5
		Tertiary	Pliocene	
			Miocene	
			Oligocene	
			Eocene —	50
			Paleocene	
	MESO-ZOIC	Cretaceous —		100
		Jurassic —		150
		Triassic —		200
	PALEOZOIC	Permian —		250
		Pennsylvanian = L. Carboniferous —		300
		Mississippian = E. Carboniferous		350
		Devonian —		400
		Silurian		
		Ordovician —		450
				500
		Cambrian —		550
	PROT-ERO-ZOIC	—		600
		Metazoans		
CRYPTOZOIC	ARCHEOZOIC	Fungi —		2300
		Blue-green Algae —		2700
		First Traces of Life (Bacteria) —		3300
		Origin of the Earth		

Impact of the Ice Age

The first major cause of ecological unsettlement, then, operative over the past 20,000 years, has been the climatic change associated with the recession of the last ice age (or glacial epoch as it is more correctly called). This led to fundamental changes of the environment and wide-scale invasion of habitats by strange life forms, and a mixing of life forms unaccustomed to each other. The second and more important factor was the change in the behaviour patterns of man, which were also promoted by this same climatic change. Let us see first how the recession of the ice has affected world ecology.

In Table 3 we can see a geological calendar of earth history. The period with which we are concerned is that known as the Cenozoic era, during which the main groups of mammals were evolved and became dominant over some 50 million years. During this time, the climate was warm and temperate; some 3 million years ago at the end of the Pliocene epoch, the last great ice age struck the earth. The history of the Pleistocene glaciations is shown in Table 4, related to the Palaeolithic stone cultures of man. The recession of the last glaciation has been plotted with a great deal of accuracy in Scandinavian countries, by recovering plant pollens from peat bogs and ascertaining the vegetational types present at different ages. This is shown in Table 4 which shows the world wide distribution of climates during the past 600 million years. The effect of the ice recession on the distribution of plants is sufficiently obvious from these tables. Nevertheless, these tables do not give a full picture of the extraordinary changes which occurred during the ice ages and their periodical recessions. The ice age deposits in Britain are revealed by the well-known boulder clays, which occur north of a line from the Thames to the Severn. In the lowest of these boulder clays are found the relics of typically ice-age animals, including extinct species of cold-loving elephants, and the so-called sabre-tooth tiger. These are overlain by riverine gravels indicating a climate probably warmer than today, and in these gravels are found the remains of animals such as the hippopotamus and straight-tusked elephants, which could only survive in warm climates. These gravels are again overlain by boulder clays containing the remains of cold climate animals, such as mammoth and woolly rhinoceros. These are again overlain by gravels containing the remains of hippopotamus, and above them more boulder clays, containing mammoth, reindeer, musk ox and arctic voles.

These changes have imposed severe strains on the life forms dependent on different types of climate. They have established themselves only to be again replaced by a different fauna; the new fauna in its turn has again been eliminated. It is hardly surprising that a goodly proportion of mammalian species have become extinct during these ages.

Table 4 *Climatic History of the Continents, Temperatures through the Ages (after Schwarzbach, 1963)*

	Arctic	North America	Europe	Asia	Australia	Africa	South America	Antarctica
Quaternary	Glaciations	Extensive glaciations	Extensive glaciations	Extensive glaciation in places	Glaciations in South	Pluvials	Glaciation especially in South	Glaciation
Tertiary and Mesozoic	Temperate to warm	Warm, arid in places	Warm, arid in places	Warm, arid in places	Warm	Warm, partly arid	Warm, partly arid	Warm at times
Younger Paleozoic	Warm,	Warm, partly wet, partly arid	Warm, partly wet, partly arid	Partly wet, glaciations in India	Glaciations	Glaciations in South	Glaciations	?
Older Paleozoic	Warm	Warm	Warm	Partly warm	Partly warm	Cool at times at least in S.	Partly cool	?
Eo-Cambrian	Glaciations	Glaciations in East	Glaciations in North	Glaciations in East	Glaciations	Glaciations in South	Glaciations?	?

Of man alone, there is an unbroken record. He gradually improved his implements of stone and his hunting methods until the end of the last great glaciation, a matter of rather recent history.

What was happening to the fauna and to man in particular in more southern areas is less clear. However, it is known that during the northern ice ages, there was a great increase in rainfall in the tropics, known as Pluvials. These had two main effects. First, the great deserts, such as the Sahara and Kalahari, were well watered and fertile, supporting savannah-type vegetation and large populations of antelopes and gazelles. Within historical times the range of the African elephant, the hippopotamus and such animals as crocodiles extended to the southern shores of the Mediterranean. In the Sahara itself are numerous rock paintings of man hunting and herding where life today cannot be supported. Great tracts of the Sahara where no rain falls today are still overlain by a layer of fertile humus from plants that have long since disappeared. Secondly, the great equatorial forests extended far to the north of their present limits. The advance of the desert has undoubtedly been accelerated by man's pastoral activities and in particular the introduction of goats. Goats — and to a lesser extent cattle — browse on the vegetation and eat the tree seedlings, making it impossible for the vegetation to maintain a precarious foothold.

Man himself, of course, had trodden the hominoid or human path long before the ice descended, though the glacial climate was to influence his development in many and important ways. We shall in the next chapter trace the ways in which climatic changes have influenced the development of man and his way of life, and have led to environmental situations which affect the earth today. Meanwhile, an attempt has been made in this chapter to illustrate the dependence of life forms on each other and of all on the habitat. In spite of this, man's ancestors defied the ice ages when the other major components of the habitat were altered or eliminated, and even in those early days showed his capacity to defy the laws of nature, and survive, a capacity which he was to exploit in later ages with such profound effects.

NOTES

1 R.N.T -W - Fiennes, *Man, Nature and Disease,* Weidenfeld & Nicolson, London 1964
2 R.N.T -W - Fiennes, *Zoonoses of Primates,* Weidenfeld & Nicholson, London, 1967
3 F. Fenner and F.N. Ratcliffe, *Myxomatosis,* University Press, Cambridge, 1965

Further Reading

Fiennes, R.N.T.-W.-, 'The Evolution and Colonisation of Habitats', in R.N.T -W - Fiennes, *Biology of Nutrition,* Pergamon Press, Oxford, 1972
Healey, I.N., 'The Habitat, the Community and the Niche', in Fiennes, op.cit.

7 THE PROGRESSION OF LIFE AND THE EMERGENCE OF MAN

From Monkey to Man

The emergence of man as the dominant species of higher life on earth is a strange story depending on an ecological paradox. The environment made man, but subsequently man drastically altered the environment. The environment shaped man's ancestors as arboreal species, but it was environmental changes which deprived man's ancestors of their arboreal habitat, forcing them to a secondary adaptation to life on the ground.

During the great biological upheaval that occurred during the Eocene/Oligocene epochs, the main groups of mammals were differentiated much as they exist today. The carnivores were differentiated from the herbivores, though they both started from a common ancestor. At this time too, the ancestors of the primates took to the trees. Thick tropical forests encircled the earth on both sides of the Equator and a rich living (an unoccupied 'niche') was there for the taking by any animal group which became adapted to arboreal living. The primates lived – as many still do – on the abundant insect life, but also fed on fruits and nuts, lizards, small birds and eggs stolen from nests. Apart from a few snakes and eagles, there were no important predators, provided that the monkeys stayed in the trees. They became differentiated into genera and species, some living in the thicker central forests, some in the lower trees of the savannahs, some in the upper forest canopies, and some at lower levels.

Perhaps life in this idyllic environment was the origin of the Garden of Eden: certainly the serpent was the most important enemy. However, ecological laws as always dictated certain evolutionary and adaptive changes. Powerful teeth were not essential to defence, and the denizens of the jungle were left relatively defenceless. The sense of smell was less important than that of sight, so that monkeys developed keen sight at the expense of scent. The orbits were enlarged to accommodate larger eyes and came closer together for binocular vision. Manual dexterity was of importance for gripping the boughs and for picking and manipulating the food. Advantages lay in a troop moving in unison, so that social habits became developed and vocal communication was advanced, enabling the members of the troop to recognise each other, and even to pass messages when danger threatened or a good feeding ground was found. All these functions led to a greater development of that part of the brain, the cerebral cortex, which is associated with intelligence. The sense of sight in particular is intimately associated with this part of the brain, whereas the sense of smell is not.

Evolution occurred also in the ways in which monkeys moved through the trees. The most primitive forms, prosimians like the bushbabies, move by jumping from one vertical stem to another. The true monkeys, as opposed to the apes, jump on all fours from one horizontal branch to another. However, the true apes – the gibbons, orang utans, gorillas, and chimpanzees – developed a new mode of progression, known as 'brachiation'. They swing, Tarzan-like, at arm's length from one branch to another, using the resilience of the branches to assist the movement and quickly transferring their weight and balance. Where possible, the feet are used too, so that the weight is distributed amongst two or more branches instead of one. This mode of progression is far more efficient and lighter branches can support a greater weight of animal. Thus we find amongst the apes massive animals such as gorillas, chimpanzees and orang utans, whereas most of the true monkeys which still live in the trees are quite small. It is interesting that monkeys of the colobus group have also learned to brachiate; these animals live solely on vegetable fodder and carry round with them a heavy digestive apparatus, so that the better weight distribution when moving in the forest is advantageous to them.

In general, primates have adhered stubbornly to their ancestral habitats, where most of them are still to be found to this day. When the forest habitat has receded due to climatic change, the monkey populations have perished. Man's closest relatives, gorillas, chimpanzees, and orang utans, still stay in their forests, although their numbers are much reduced as forests are removed under human influence. Two groups of monkeys have to a large extent forsaken the forest habitat, the baboons and the patas monkeys. Both groups show modifications of structure and behaviour which have enabled them to survive away from the forest. The teeth are much enlarged for defence, especially in the males, and troops have improved social organisation adapted to the protection of the troop as a whole. The intelligence of baboons seems to be of a higher order than that of other monkeys, while the patas have developed long limbs and can move rapidly over ground surfaces. Their feeding habits have changed also, and they have learned to grub in the ground for roots, and the baboons sometimes hunt and kill animal food. Some Asian monkeys also spend a portion of their time on the ground and some have learned to find food in rivers and estuaries. So there are modern parallels for a phase of existence through which the human stock must have passed.

It has been proposed, particularly by the Dutch primatologist Dr A. Kortlandt[1] in his 'dehumanisation' theory, that surviving chimpanzees fall into two groups, those that are partially ground-living, and those that live solely in the trees. He believes that the ground-living groups may be related to proto-human stock, which began to be adapted to ground-living but reverted to the ancestral mode of life, whereas the proto-human stock stayed on the ground. He has shown by various experiments in Africa that the partially ground-living chimpanzees more readily use sticks as clubs and weapons and are more naturally aggressive to potential attackers.[2] Appearances apart, chimpanzees are so closely related to man judged by anatomical and many other criteria as to make this feasible. Moreover, as has been shown by Jane Goodall,[3] savannah chimpanzees hunt and devour antelopes, bushpigs, monkeys and domestic fowls, whereas forest chimpanzees are strictly vegetarian.

However that may be, it is certain that no group of apes or monkeys has deserted tree living for the ground unless forced to do so by ecological circumstances, that is because the climate changed and the forest habitat disappeared. So it must have been with man, and palaeontological evidence tells us that this was so. However, unlike the baboons and patas monkeys, man overcame the dangers and feeding difficulties of life on the ground in a different way. He did not develop the dangerous teeth of the baboons and patas. Instead, he used and subsequently made weapons of offence. He used sticks and stones and bones, and presumably anything else that came to his hand. He developed an improved social organisation, and perhaps most important of all he learned to communicate on a far higher plane, eventually developing language and speech. This demanded an increase in the size of the brain, improved powers of thought and eventually the ability to communicate thought and abstract ideas.

The present-day distribution of the great apes, with gorillas and chimpanzees in Africa and orang utans in Asia, shows that the great ape stock must have migrated between these far distant points at a time when the earth was humid and the great African forests were linked with the Asian. Man is supposed to have descended from a group of apes which lived in Miocene times; fossils of one type of these have been found in the Siwalik Hills in India, and it is known as *Ramapithecus;* the other was found in Kenya and given the name of *Kenyapithecus.* However, *Kenyapithecus* so closely resembles *Ramapithecus* that both are now included in the single genus of *Ramapithecus.* Man was presumably descended from the Kenya branch of his family and – as Kortlandt believes – from some form close to the chimpanzees. Claimants to human ancestry have been found also in lower Pliocene strata, but the fossil record from the upper Pliocene is deficient. However, in the lower (early) Pleistocene there emerges in south and east Africa a group of creatures known as *Australopithecus,* whose resemblance both to *Ramapithecus* and to man is very close. Some forms of *Australopithecus* are said by some to merge into the earliest fossils that can be definitely classified as *Homo, H. habilis. H. habilis* merges into a form of man, *H. erectus,* who was widespread from Africa to the Far East, and who is regarded as the direct ancestor of *Homo sapiens.* Both *Australopithecus* and *Homo habilis* are believed to have fashioned and used simple pebble tools. One group of *Australopithecus, A. robustus,* developed massive jaws and teeth, believed to have been associated with a hard vegetable diet. The less specialised lived by hunting small antelopes and other animals, whose bones they also used extensively as weapons. They also attacked and killed their own people and possibly indulged in cannibalism.

However, the position of *Australopithecus* in human ancestry has never been universally accepted since some eminent students of the subject maintain that anatomically he was too ape-like to be placed in the direct line of descent. Recently, discoveries by Richard Leakey in the Lake Rudolph area of Kenya lend support to those who hold this view. A fossil form of *Homo* with definite human characters and a larger brain case than *Australopithecus* has been found of a date – some three million years ago – which is older than that of most australopithecine remains. This discovery would suggest a common ancestry for *Homo* and *Australopithecus,* the latter being unsuccessful and becoming extinct.

The emergence of man from the apes is one of the most fascinating stories of ecology. In spite of close anatomical and phylogenetic resemblances, let us be clear about the great gulf which separates habitat-independent man from his cousins the apes, so habitat-dependent as to be in danger of extinction because of the decline of the forests. Anatomically man is distinguished by his upright gait, his manual adaptation and the size of his brain. In powers of thought and communication he is as far removed from the apes as the apes are from the dinosaurs. In spite of this, man's special powers must have been evolved over a very short period indeed, probably less than a million years. This development could not have occurred without a period of intensive selection, in which enormous numbers of the proto-human race failed to survive. In virtue of this, a race of man had emerged in northern areas during the Pleistocene, indistinguishable from ourselves and evidently with the capacity to invent and learn the higher technologies which are a feature of our culture. How did they get there and how did they live?

Early Human Society

As we have seen, the earliest apes which can claim human ancestry were present on earth in as widely separated habitats as Africa and India during the late Miocene and early Pliocene, when the climate was warm and temperate. Little is known of the upper (late) Pliocene. However, during the Pleistocene when most of Europe, Asia and North America were glaciated, human creatures appeared to have moved north into the glaciated areas and to have developed a highly satisfactory and profitable way of life. For a creature developed in the warm climate of Africa and having no natural defence against the cold, this was indeed a strange thing to do. It was plainly possible only because of high intelligence, and the ability to protect himself against the cold by making clothes and footwear and as a result of the discovery of fire. His hunting ventures too required skill and organisation, since he was hunting animals of considerable size such as mammoths and wild horses, and he needed to protect himself against large and dangerous predators, such as sabre-tooth tigers and the large arctic hyaenas.

In the earlier years of this era, man is best known from the stone implements he manufactured, of increasing sophistication as the years passed by. As the successions passed from the boulder clays of the glaciation to the gravels of the warmer interstadials, it becomes clear that man at this time was more numerous during the periods of cold. During this long period of 100,000 years, the tribes of the north evidently became accustomed to the arctic climate and this provided for them a habitat with which they became well associated.

The main type of vegetation in these areas was that called 'tundra'. This was an arctic or alpine form of vegetation which was low growing. It was very lush because of the numerous streams emanating from the ice cap which permitted rapid growth during the summer season. It supported enormous herds of grazing animals, which owing to the open nature of the country were easily hunted by an animal that had developed the skills of man. Indeed, Palaeolithic man constantly indulged in acts of slaughter, which to say the least of it were improvident, and he has been accused of being responsible for the extinction of the mammoth. Men appeared at that time to live in small groups. The main habitats were caves or rude huts built of mammoth bones and tusks, which were to the south of the ice fringes. The animals which were hunted were to a great extent waylaid on their migration routes as they moved north in summer and south in winter. Evidence suggests that parties of young men would travel north in the summer months and set up hunting camps by the sides of lakes. They would secure a bag of deer and others of the larger animals, and birds such as geese, and return to base with their provender when winter came. The meat could then be preserved beneath the snow or in whatever other type of deep freeze nature provided. They probably used sledges made of skins, and indeed skins and furs were much used for warm clothing.

By the end of this period, perhaps 20,000 years ago, men of modern type (the so-called Cro-Magnon race) had appeared in northern regions and had been living there in this kind of way for 100,000 years or so. No doubt we would still be living there in the same way, but for the climatic change which ended the Ice Age. Evidently during this time genetic forces had discriminated in favour of mental development. The more intelligent, at any rate, were equipped with the same mental powers as has enabled man to develop the technology of the space age. It would be inconceivable that these powers could have been developed in so short a time if it were not so. In art, too, the vigorous paintings of the well-known caves are in no way inferior to modern art, though less sophisticated. It is also evident that religion and magic played a great part in their lives. However, until the Ice Age ended, there was no incentive to create a change in a thoroughly satisfactory way of life. This way of life, however, was associated with the easy hunting provided by the open tundra. If the country became less open, life might become more difficult.

Just such a change occurred with the recession of the ice. As we have already seen, with this recession the country became forested. First pines invaded, then birch, and finally mixed forest. Man was not equipped to hunt in this kind of terrain, and had he been any other animal, he would have disappeared with the change of habitat. Some, of course, migrated further north, the ancestors of the Eskimos, and continued to live by hunting reindeer and seals. During the Mesolithic period, the Caucasian types who stayed behind had a very difficult time. In Scandinavia, their settlements – now permanent and not temporary – are found in the open land between forest and sea. Vast middens on the outskirts contain the relics of what they ate. There was much shellfish, and the bones of such forest animals as they could hunt or trap; they also gleaned nuts and other vegetable produce in the forest. The forests were held in awe and no doubt many legends and myths about the gods of the forests arose at this time. It is evident, however, that man was a shadow of his former self. His vigour had diminished and he was unable to recover in the face of this drastic climatic change. It was not to be expected that so resilient and intelligent a creature could be permanently subdued by these unfavourable circumstances, and the wind of change blew strongly over the land, a wind that was to bring profound ecological consequences to the whole face of the earth.

In the Near East, the Mesolithic period only lasted a few hundred years. In northern Europe it dragged on over 2,000 to 3,000 years. Its end was heralded by three revolutionary developments. The northern peoples domesticated the wolf, whose keen scent enabled it to hunt in the forests where man's keen eyesight was of little use. The fashion spread quickly to other parts of the world, and domestication of ox, pig, goat, sheep and horse followed. The second important development, which may have occurred in central Europe was the invention of the polished stone axe. These were long, narrow axes made, as the name implies, from hard polished stone, and well

Table 5 *The End of The Ice Age – climatic and vegetational fluctuations in Scandinavia since Palaeolithic times*

Glacial stage	Date	Climate	Forest zones	Archaeology
Post-glacial	400 B.C.	Sub-atlantic (cold, wet)	Beech	Early Iron Age
		Sub-boreal (warm, dry)	Mixed oak forest	Bronze Age Neolithic
	2,500 B.C.	Atlantic (warm, moist, oceanic)	Mixed oak forest	
	5,000 B.C.	Boreal (warm summers, cold winters; dry, continental)	Pine and hazel	Mesolithic
	6,800 B.C. 7,900 B.C.	Pre-boreal (transitional)	Birch, pine, willow	
Late-glacial	9,000 B.C.	Sub-arctic	Tundra	
		Warm oscillation	Birch and pine	
		Sub-arctic	Tundra	Upper Palaeolithic

hafted with wood. By means of these axes, the forest trees could be cut, and the forests no longer remained the masters. Not only this, but in the succeeding Neolithic era a whole range of carpenter's tools were developed from stone. There were saws made by fixing flint teeth in wooden frames, and adzes, even stone planes. For the first time wood could be worked. One of the most important developments from this was the building of river boats and ships that could go to sea, and the ground was prepared for the development of trade and commerce.

Meanwhile, the Near Eastern peoples had discovered that some

Fig 26 STONE AXE. Reconstructed by mounting the Neolithic flint head on a copy of a Neolithic haft preserved at the bottom of a bog. It was found that the full swing of the modern woodsman often chipped or broke the head. Using short, rapid strokes, the experimenters learned to fell trees more than a foot in diameter in thirty minutes. To fell small trees they chopped all the way around the trunk.

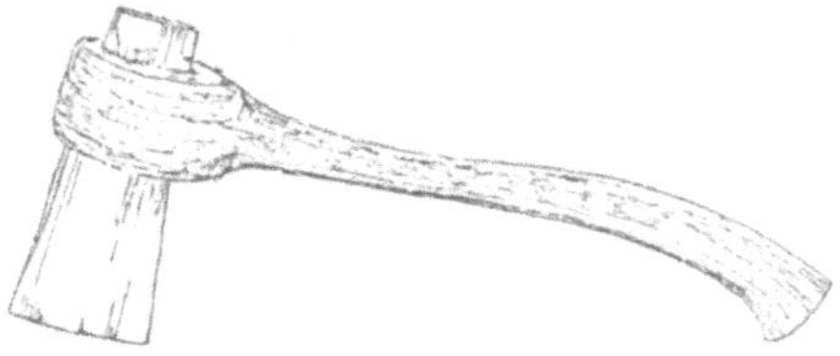

of the seeds they gleaned for food in these lean times could be made to grow if the ground was cleared and dug with a 'digging stick'. Particularly good results were obtained with the seeds planted in an area where the cut brushwood had been burnt. They took to clearing areas of woodland, burning the wood, digging and planting. Barley and wheat were the first crops to be planted. This primitive treatment of the soil necessitated frequent moves to new sites, and this form of shifting agriculture moved slowly across Europe until it eventually reached our northern friends with their domestic dogs. Meanwhile, peoples living along the great river deltas found that these seeds grew particularly well in the mud deposited during the annual floods. So developed the great river-based civilisations on the Nile, the Tigris and Euphrates, the Indus and the Yellow River in China. From these developments came the city states, commerce, banking, war, empires, and so to the recorded history of man.

While all this was happening, there was a group of peoples living in the vast steppe lands of eastern Europe and central Asia. These areas had not been invaded by forest; the chernozem soils supported a rich growth of valuable grazing grass, of which the dominant species is known as *Poa pratensis;* this grass was introduced to North America and is the famous Kentucky blue grass. This was all horse country and the wild tarpan horses were present in vast herds; in the colder north-eastern parts of this area in Siberia were the Mongolian wild horses (Przewalsky's Horses). The human tribes in this area were less affected by the change of climate and environment. They still lived in open terrain, and could pursue their traditional lives as nomads and hunters. They had been hunting wild horses from time immemorial for food. Although they were nomads, they were not to be left behind in the march of progress, and they were to develop in a way which has had and still has a profound influence on history. Their first contribution was to domesticate the horse and learn to ride it with the help of a crude bridle and bone bit, but no stirrups. They then turned to carpentry, invented the wheel, and made waggons, some of which were drawn by oxen and some by horses. Their crowning achievement was the construction of the two-wheeled, horse-drawn chariot, with light spoked wheels. They thus provided themselves with the two major military weapons which were to be supreme for centuries to come, cavalry and chariots. Nor did they hesitate to use these weapons; they continually invaded their richer neighbours to the south and east, and were a constant menace throughout the Seleucid and Roman Empires, up until the invasions from the east of Genghis Khan in the fourteenth century A.D.

It was in the late Neolithic era that commerce was really opened up through the then known world. The horse riders of the steppes took to the water as if born to it, and began to build excellent wooden ships, with which they and their emigré descendants traded throughout the Mediterranean, by way of the central European river systems (the 'Amber Route'), down the east coast of Europe from the Baltic into the Mediterranean, to the mouth of the Indus in India and down the east coast of Africa. There resulted a thousand years − 3,000 to 2,000 B.C. − of trade and prosperity in a climate that had so improved as to be sunnier and warmer than it is today. The evils of excess population growth, crowding, rivalry, greed, and continual warfare, were to follow in the succeeding Copper, Bronze and Iron Ages.[4]

The breathless bound of *Homo sapiens* from Ice Age nomads to agriculturists and traders and thence to the space age has all occurred within a few thousand years, perhaps six thousand. During the Ice Age, however, man no less than other groups of animals had been subject to the isolationist forces proposed by Alfred Russell Wallace. During the 100,000 years or so of this isolation, the races of man had diverged and showed clear racial characteristics developed in conformity with the habitats in which they lived. The races had diverged in physical characteristics such as colour, and in organising ability and intelligence. The peoples of the north had advanced furthest in technology; the darker skinned peoples of the south were left behind in this respect. To this day, peoples exist who are − or until very recently were − still living in a stone age culture, such as the Bushmen of South Africa, the Australian aborigines, peoples of New Guinea, and enclaves of peoples throughout Indonesia and the Indian subcontinent. To these we could add the Eskimos and the American Indians. The mixing of and intercourse between these races pose problems of the greatest difficulty in human ecology. As so often in ecology, it is the meek that inherit the earth. The apes that left the trees developed into man, and the human races forced into the bitter climate of the Ice Age became the dominant races of the earth. In the event of a total breakdown of western civilisation, who can say from what branch of the human race the next upsurge would come?

NOTES

1 A. Kortlandt, 'How do Chimpanzees use Weapons when Fighting Leopards?', *Yearbook* of the American Philosophical Society, 1965, pp. 327-32
2 A. Kortlandt and M. Kooij, 'Protohominid Behaviour in Primates (Preliminary Communication)', *Symp. Zool. Soc. London,* 10, pp.61-88 (1963)
3 J. Goodall, 'Feeding Behaviour of Wild Chimpanzees (A Preliminary Report)', *Symp. Zool. Soc. London,* 10, pp. 39-47 (1963)
4 G. Bibby, *Four Thousand Years Ago,* Collins, London, 1962

Further Reading

Bibby, G., *The Testimony of the Spade,* Fontana (Collins), London, 1962
Carrington, R., *A Million Years of Man,* Weidenfeld & Nicolson, London, 1963
Chiarelli, B., *Taxonomy and Phylogeny of Old World Primates, with References to the Origin of Man,* Rosenberg & Sellier, Torino, 1968
Cloudsley-Thompson, J., *The Zoology of Tropical Africa,* Weidenfeld & Nicolson, London, 1969
Fiennes, R.N.T -W - and Fiennes, A., *The Natural History of the Dog,* Weidenfeld & Nicolson, London 1968, and Natural History Press and Bonanza Books, N.Y., 1969
Goodall, Jane, *My Friends, the Wild Chimpanzees,* National Geographic Society, Washington, 1967
Goodall, Jane Lawick-, *In the Shadow of Man,* Collins, London, 1971
Napier, P., *Monkeys and Apes,* Hamlyn, London, 1970
Weiner, J., *A Natural History of Man,* Weidenfeld & Nicolson, London

8   MAN AND EARTH ECOLOGY

Man then is an arboreal animal, secondarily adapted to living on the ground. In this situation, he has also become adapted to be a highly organised social nomadic hunter in open situations such as tundra or prairies. By a third adaptation, he became a crops/pastoral agriculturist, by which many of the species still earn their living. By a fourth adaptation, he has become a highly urbanised animal, living in conditions of high population not matched elsewhere in the animal kingdom, except on a much smaller scale with ants, bees and termites.

As a predator during the Ice Age, he was a menance to his prey and other animals. If opportunity offered, he killed in excess of his needs – and still does. He selected for the kill the best animals in the herd, instead of the sick and the laggards usually taken by other predators such as wolves. Furthermore, he hunted not only for food, but also for skins and furs required to keep him warm, and in historical times at any rate he has hunted purely for sport and pleasure. He has been blamed for the extinction of the mammoth, of which he certainly killed enormous numbers. Even so, his responsibility for this has been questioned and many believe that the disappearance of the mammoth is to be attributed rather to sudden climatic change. While many other species became extinct at times during the Ice Age, this too is likely to be associated more with climatic change than with human depredation. Human numbers appear to have been quite small and as nomads the bands would be likely to shift their hunting grounds, and it is difficult to see how they could have exterminated whole species of animals.

At the end of the Ice Age, the position was radically changed. Animals that had previously been hunted were brought into captivity and reared as domestic stock. Conditions for the wild forebears of these animals deteriorated with the enclosure of land for farming or ranching, and because they were deliberately destroyed as being supposedly dangerous, or because they competed for pasture or stole the crops or killed the domestic animals. From Britain there have disappeared wolf, bear, beaver, polecat, aurochs, horse, elk, reindeer, and birds such as ospreys and golden eagles. Other species are endangered, and all over Europe, Asia, and North America the same story can be told. The rich fauna of Africa is threatened.

Vegetation-wise, the ability to fell the forests which started with the invention of the polished stone axe has had profound effects. Man's demand for timber, for building houses and ships, for making paper, and for burning, has been insatiable. Trees have been cut without adequate replanting programmes. They have been removed from vulnerable sites, such as watersheds, giving rise to diminution of rainfall, flooding, erosion and lowering the water table. Furthermore, animals inimical to the habitat have been introduced, such as rabbits and goats; tree seedlings have been cropped by them, so that natural regeneration has been prevented. All of the trouble cannot be laid at the door of modern man. Even in early dynastic times in Egypt the depredation of marginal but still useful lands had started. Plato lamented the disappearance of the trees from the hills of Attica, and the effect this had on water supplies. These hills are still denuded, and so have been the hills of Palestine since biblical times.

In the early days of shifting agriculture, probably little harm was done because of the necessity to leave the land to regenerate after a few seasons. In temperate regions too, well-planned agriculture, as practised in Celtic and Roman times and in countries such as Britain until recently, appears to have had no lasting ill-effects on the soil. Until recently too, the practice of shifting agriculture in tropical countries such as Africa has probably not been permanently deleterious. However, intensive agriculture in the tropics can easily lead to permanent loss of farm land unless very great care is taken. This occurs for two reasons. First, heavy tropical storms can easily remove thin topsoil which has taken centuries to be deposited (sheet erosion); they also cause great gullies which may be ten or twelve feet deep (gully erosion). Secondly, the effects of the hot tropical sun on the mineral-rich subsoil leads to 'laterisation', by which the soil is compounded into rock.

In so far as farming leads to monocultures, whether of crops or animals, it must also lead to the multiplication of pests, parasites and disease-causing organisms. The old-fashioned mixed farming with small fields and frequent hedges to harbour birds permitted these to be largely controlled by natural means. In intensive systems, these natural controls cannot operate, bird populations fall and pests – including bird pests such as pigeons – multiply. So the fields must be treated with chemicals and by burning the stubble. By this means, the soil-living ecosystem is also destroyed and such creatures as voles are deprived of their food and disappear. The ecological effects are profound. Already, wind erosion is threatening the soils of Norfolk in eastern England because of the removal of the hedges, and nobody knows what will be the long-term effects on soils deprived of their crumb structure and fed with massive top dressings of artificial fertiliser.

For North America, the Industrial Revolution gave rise to widespread erosion problems. The momentum of the revolution could only be sustained by large production of grain to feed the increasing factory population. Thus grain farming replaced cattle farming over huge tracts of the Middle West; the land was seriously mined and lost its fertility; dustbowl conditions were created. For England, too, this development was a tragedy, since the availability of imported grain permitted too rapid a growth of population and urbanisation at a time when little was known of hygiene and disease control, and large tracts of good countryside were destroyed. Aerial and water-borne pollution also started at this time. The effects are felt to this day.

Population Growth and Urbanisation

We can here take a quick look at the problems of population growth and urbanisation. We have already studied certain basic principles, such as:

1. That populations are maintained at optimum levels demanded by the ecosystem;

2. This maintenance is achieved by production of excess young and elimination of those least fitted to survive;

3. That mating and parenthood are denied to the young of the species until they have acquired a 'territory', that is a sufficient area in which to raise and feed young and provide protection;

4. That, if the population becomes excessive either because of a breeding rate that is too high or because conditions become unfavourable, certain glandular mechanisms operate to induce a state of 'stress', during which the breeding urge and fertility become diminished, aggressiveness towards other members of the community is enhanced, and susceptibility to disease is increased.

The most primitive human communities, such as the Kalahari bushmen and the Australian aborigines, were, and to a certain extent probably still are, subject to these natural laws. So, we must suppose, were all human tribes up to the time of the agricultural revolution. In subsequent human history, these natural laws have been progressively put aside, it would appear by deliberate intent, until in advanced human communities they are virtually inoperative, except in certain ways discussed below. However, let us examine these principles in turn in relation to human ecology.

The first principle, that populations are maintained at optimum levels for the ecosystem, has plainly been defied. Since the invention of agriculture, man has deliberately manipulated the environment and refused to conform to it. In many areas and at many times, he has exploited his habitat and caused permanent deterioration, to such an extent that it can no longer support his agriculture. When the damage is very extensive and deserts are formed, of course the human population disappears; where the damage falls short of this, the population tends to persist without great diminution of numbers at subsistence or below subsistence levels, malnourished and living in abject poverty. Disease is rife and frequent epidemics occur; famines and floods from destruction of ground cover recur at intervals. Still, the population maintains a high birth rate. However, expectation of life may be no more than thirty years, and infant mortality in particular is fantastically high. It is difficult to understand just why natural population-controlling mechanisms are not operative. Two possible causes may be suggested. First, such situations are not entirely unknown under natural conditions. Man could be regarded as an introduced species to a new environment. Under such circumstances, it often happens that the introduced species multiplies vigorously and destroys the environment, as has occurred with rabbits in Australia. The same has been happening with free-ranging animals, such as elephants, confined in game parks in Africa; their numbers increase and the confined habitat deteriorates. The problem is partly associated with the lack of predators in the new environment which will serve to limit the numbers of the introduced species. For whatever reason, man has developed to some extent into his own predator and limited numbers by wars of aggression, human sacrifice, cannibalism, and so on. The second cause may be associated with a deliberate reaction of an intelligent, reasoning creature in the face of a high mortality that threatens his race. Fertility cults are a feature of all religions, and exaggerated importance has been attached to an early and high rate of breeding. Fertility cults evidently first emerged in prehistoric, Palaeolithic times. No doubt in those times life was hazardous and high fertility was necessary to maintain the numbers of the community. Since man adopted a settled existence, the high fertility has been retained in spite of overpopulation and in defiance of nature's laws with resulting low standards of life and periodical mass exterminations from natural causes.

The second principle states overproduction of young and elimination of the least fit. Population depends on the balance between young produced and rate of survival. Animals that are especially vulnerable produce large numbers of young and survival is low. Small birds produce large clutches of eggs and less than half come to maturity or themselves breed. The extreme of this is seen with parasites, the larvae of which must find a new host, a somewhat fortuitous matter. A female tick produces thousands of eggs which develop into larvae, but the general tick population under normal circumstances does not increase. The same is true of intestinal worms and of protozoan parasites carried by ticks, and other vectors. The tendency of agricultural man also is to overproduce, but the balance is not redressed by comparable casualties. Epidemics and catastrophes apart, the main population-limiting factor in human history has been chronic ill health. This operates to the greatest extent on the very young and on the young and the mother at childbirth. In primitive, crowded communities neonatal deaths and deaths of mothers at childbirth are very high.

Until weaning, newborn children are protected to a great extent against diseases by immunity acquired from the mother in the milk. Thereafter, they are exposed to numerous diseases, of which malaria is probably the most important. A significant proportion of the juvenile population (up to 20 per cent) may succumb to this and other diseases, such as intestinal infections, pneumonia, and so on. If the population is malnourished, as is usually the case, the effect of disease is intensified and deaths may occur also from nutritional diseases, such as kwashiorkor due chiefly to lack of protein. Furthermore, intelligence and general performance during adult life will be impaired. Nevertheless, the birth rate remains at a level high enough to perpetuate these conditions, in spite of an early death rate in adult life, from diseases such as cholera and tuberculosis. The impact of more developed civilisation on peoples living in such conditions is disastrous. Attempts are made to improve food production and to reclaim exhausted land, but the impact of medical science is far more rapid in its effect. The killer diseases are brought under control and childbirth is made more safe. The result is population explosion on an unprecedented scale, with an exacerbation of pre-existing problems. Cataclysms and catastrophes affecting adult people become more severe. These are met by international relief programmes and shipments of surplus food from overseas.

The third principle that the young do not mate and breed until they have acquired a territory is no longer a rule of human ecology, though territorial instincts are strongly operative in us. Territory can belong either to a pair, as with a pair of birds which claim the right to the worm on the front lawn, or to a group of animals living in social conditions. Man appears to possess both instincts. A tribe will fiercely defend its own territory from invaders. Teenagers in cities, who form themselves into bands, seem to demarcate territories belonging to the different bands, and they usually respect each other's territories. On the other hand, it is the ambition of each young couple on marriage to own their own house, however humble, in which they can live together and raise their families. This instinct could be the saving grace of urbanised peoples, and many couples do obey the ancient instinct and defer marriage or babies until they have acquired their own home. This is made the more easy for them in Western societies, where birth control methods are readily available, and where the urgency of sexual instincts can be satisfied without the danger of unwanted children. However, this has not always been the case and is still not the case amongst peoples at a less advanced level, though evidence suggests that in many cases means of birth control would be welcomed. Amongst many peoples, the acquisition of a territory for young people is a virtual impossibility, but this does not act as a deterrent to breeding. Nevertheless, the instinct still exists and is a powerful force which may yet be important in human ecology in the future.

The fourth principle concerns the operation of 'stress' mechanisms in response to overcrowding. This is a somewhat complex problem, and will require rather more detailed study. Stress undoubtedly plays an important part in human ecology, although the human race as a whole resists stressing situations which would be intolerable for many wild animal species. Hans Selye, who has been a pioneer of the study and concepts of stress, regards stress as a general reaction to trauma in the same way as inflammation is a local reaction. The 'stressors' which cause the stress reaction can be either physical or psychological. In the present context, we are concerned with the psychological aspects, which have been studied extensively by animal ecologists in relation to population problems. The essential mechanism is fairly simple. The important pituitary gland acts as a coordinator of endocrine activity, by producing hormones that influence the other endocrine glands including the sex glands. It is situated at the base of the brain and its own activity is controlled by a nearby portion of the brain. When an animal is stressed, the pituitary sends a hormone to the adrenal gland, which stimulates its activity. In response, the adrenal gland produces more of its hormones which counter the stress situation, while the sex glands (and sex urges) are diminished. The results are a diminished urge to mate and reduced fertility – even abortion – while the animal enters a state of nervous tension, adopts unnatural habits and behaviour, and becomes more aggressive. If, as a result of continued or repeated stresses, the adrenal glands become exhausted, the animal will die since its secretions are essential to life.

The extent to which these stressing mechanisms arising from a crowded existence are operative in human ecology is a matter which must be considered. While man reacts atypically, the evidence is suggestive that urbanised human communities are indeed subject to the pressures of stress. At some time in their lives 10 to 20 per cent of men and women in urban areas receive treatment for neuroses of some kind, and an unreasonably high proportion of the population is permanently incapacitated or hospitalised for mental defects of one kind or another. It is unlikely in the extreme that any wild community, or primitive human community, could or would support so high a proportion of ill members. They would almost certainly fall prey to the natural predators of the species. However, man today has no natural predators, except his disease-causing parasites, and these, followed the many years of impotence against them, he more or less successfully controls. His principles are such that he preserves the weak and the unfit; and the drop-outs are, therefore, preserved as a burden on the community as a whole.

It might be thought that the long-term effect of this would be to put evolution and natural selection in reverse with a consequent genetic weakening of the community as a whole. However, Dr. Penrose, the British geneticist who has made a profound study of the subject, does not believe that this is necessarily the case.[1] He believes that, appearances notwithstanding, evolution is powerfully at work in the human species, but in a different way operating more on the unborn foetus than post-natally. He points out that more than 50 per cent of people in urban communities are 'genetically unfit'. This is to say that either they have no children themselves, or if they have children those children do not breed. Thus, in each generation there is a better than 50 per cent selection. The figures he gives are as under:

1. *Prenatal deaths*   15%
2. *Of the remainder there are*
    a. Stillborn   3%
    b. Deaths of the newly born   2%
    c. Deaths before maturity   3%
3. *Of the survivors*
    a. Those who do not marry   20%
    b. These that remain childless   10%

The members of the population to breed most successfully are the middle-of-the-road ordinary people with a wide genetic constitution; those at the top or bottom of the brilliance league tend to be less productive. Therefore a wide genetic pool of capability is retained and those best adapted to their way of life are most likely to contribute the future generations. With so large a degree of infertility, there evidently does exist a widespread unwillingness or incapacity to breed. This, then, is just what we have seen to occur in wild animal populations that come into stress. The fact that, in spite of this, there may still be an upward movement of population numbers does not affect the argument, and it may well be that at some upper limit not yet reached a downward trend of population would result.

It can be argued then, without dogmatism, that urban crowding does indeed affect population trends amongst a section of the people. Are there other signs of stress effects to be seen in urban peoples? There are abundant signs, though it is impossible to prove that these signs are the effect of crowding or urbanisation. Let us consider them. The whole can perhaps be summed up in the term 'urban unrest', namely a feeling of insecurity and anxiety amidst the outward signs of affluence and prosperity. Here we may just list the types of unnatural behaviour that might be associated with stressing problems. First, there is the problem of young adolescents and their antisocial behaviour of whatever kind. Secondly, sexual freedoms and unnatural sexual habits. Thirdly, addiction to drugs and excess alcohol. Fourthly, the high level of lawlessness and crime.

Human history itself has been a story of alternating periods of vigour and decline. During the periods of vigour, wealth and population have increased. The periods of decline have been marked by signs of unrest leading to aggressiveness and wars, often accompanied by famine and pestilence; death and population decline follows. Here again, human history follows the ecological pattern of wildlife societies.

We can, then, discern in human society the operation of natural laws which govern pre-agricultural situations though they operate over much longer time intervals. Two new factors have been introduced into the ecological picture in recent times, namely the ability to control population increase by methods of family management, and the far-reaching advances that have been made in controlling disease. Amongst more advanced peoples, conditions of life have been so ameliorated that it is difficult to visualise what it was like to live even 100 years ago, and this amelioration has been achieved not only for the rich, but for the great mass of the people also. The expectation of life for the poor a century ago was under forty years, and even for the rich no more than fifty. Today, one can expect to live to seventy or more. Who would return to the good old days? Their dark, damp, and insanitary houses? Lack of water, drainage, sanitation? Long working hours without holidays? Deaths in infancy of up to half the children born? Twopenny gin and degradation?

Is the present the crest of the wave, before another period of decline? The general conditions of unrest would suggest it. An understanding of these problems may show the way to avoid it. Populations must be controlled at a point below that which leads to aggressiveness and instability, otherwise ecological laws will take over. Finally, the benefits of modern living must be extended to more backward peoples. This means, above all else, education, true education − not just literacy. It is useless to prolong by medical means the lives of people who are unable or unwilling to come into ecological balance with their environment.

NOTES

1 A.L. Penrose, *Biology of Mental Defect,* Sidgwick & Jackson, London, 3rd edn.,
   1963, and D.F. Roberts and G.A. Harrison (eds), *Natural Selection in Human
   Populations,* Pergamon Press, Oxford, 1969

Further Reading

Andrewartha, H.G. and Birch, E.C., *The Distribution and Abundance of Animals,*
   University of Chicago Press, Chicago, 1954

Bertram, Colin, *Adam's Brood,* Peter Davies, London, 1959

Harrison, G.A., Weiner, J.S., Tanner, J.M., and Bernicott, N.A., *Human Biology,*
   Clarendon Press, Oxford, 1964

Kormondy, E.J., *Concepts of Ecology,* Prentice-Hall, Englewood Cliffs, New
   Jersey, 1969

Lack, D., *The Natural Regulation of Animal Numbers,* Oxford University Press,
   Oxford, 1954

Selye, H., *The Story of the Adaptation Syndrome,* Aeta Inc., Montreal, 1954

Wynne-Edwards, U.C., *Animal Dispersion in Relation to Social Behaviour,*
   Oliver & Boyd, Edinburgh and London, 1962

9 THE GLOBAL ECOSYSTEM

Natural Laws of Ecology

It is a common trick in ecological studies to enclose a piece of land, protecting it from external influences and thus removing factors which may be deflecting the primary succession. In this way, one can study the regeneration processes and discover what vegetation would naturally develop. The influencing factors to be excluded may be various, such as rabbits, grazing by livestock, annual grass burning, and so on. The results of such experiments are often enlightening, and they may give useful information on the best ways to manage the land.

It is intriguing to consider what the results would be of a similar experiment, if man were totally removed from the earth. The main results can be deduced from simple observation. We all know how quickly weeds and undergrowth take over a neglected garden, until it becomes virtually impenetrable; how the ivy claws down walls and buildings, and kills introduced fruit and other trees. Even in derelict sites, such as bomb sites in a city like London, natural vegetation quickly returns. Around cities, a whole host of creatures, such as sparrows and pigeons, even foxes and hyaenas, which have learned to live by scavenging man's waste materials, would be forced to revert to more natural forms of life or disappear. Likewise, those that scavenge the grain fields would find life less easy. Within a few centuries most visible signs of human occupation are likely to have disappeared, and it would require an archaeologist of the future to rediscover and marvel at a lost civilisation.

In brief, the world's habitats would again be ruled by the natural laws of ecology. Where climate and soil favoured the development of forests, the land would revert to forest. In Britain, low-lying areas would again become covered by the thick tree growth which defeated our Palaeolithic ancestors. The moors and upland downs, where Neolithic man had his ridgeways and settlement areas, would revert to their natural vegetation. The great prairie lands of the world, such as those of the chernozem soils of North America and central Asia and Europe, would again become rich natural grazing lands. Man-made aerial pollution would disappear. Even the balance of carbon dioxide between atmosphere and ocean would be readjusted; any effects of releasing fossil carbon would be removed such as the supposed 'greenhouse' effect; and particulate matter released into the atmosphere by human activities would subside. If man's way of life has indeed affected the climate in these ways, then the climate would revert to that dictated by solar radiation, and whatever other influences determine the climate. Soil communities too would regenerate, and the removal of man-made pollution from inland waters, estuaries, and the oceans would enable the regeneration of aquatic creatures; fish would again teem in the seas, and in the inland waters.

On land surfaces, herbivorous animals would build up large populations within a short time. The bison would again cover the plains of North America. Horses would become feral, as did the mustangs in America during the last century, and take over the plains of Europe and Asia, possibly in America also. In forested areas, deer and elk would become numerous. In Africa, elephants and rhinoceroses would come into their own again, and vast herds of antelopes, gazelles and zebras would cover the plains. Predators might be slower to recover, especially in countries such as Britain where so many have been exterminated. However, the domestic dog has displayed in many instances its capacity to become feral and redevelop the properties of a social hunter. So we could expect the re-emergence of canine stocks resembling wolves, which would prey on the great herds of herbivores.

According to many of the more baneful prophets of doom, this is indeed what may happen. It is said that we are fouling our own nest to such an extent that recovery is already impossible. Some would go further and say that we will take all other living things on earth with us, destroyed by universally lethal radiation.

Some might think that man has made such a mess of the great opportunities offered him that from the earth's point of view his disappearance would be highly desirable, and that in course of time some other form of highly intelligent life would appear that would do better. However, would it? And if it did, would it not also pass through the same difficult stages that man is passing through at the present time? The point deserves some consideration both from the practical and the philosophical point of view.

Some of the more successful groups of animals are those which have become habitat-independent. Man is a case in point. Adapted to an arboreal habitat, he became secondarily adapted to life in open, terrestrial country and changed his diet from one that was mainly vegetarian to one that was partly carnivorous, and certainly omnivorous. The third refinement to appear on the ecological scene has been man's manipulation of the environment over the past 5,000 to 10,000 years, for the sake of his agriculture, and a further reversion to a primarily vegetarian diet.

These are surely inevitable ecological sequences, and should man disappear some other species will in the end appear and in its turn also attempt to manipulate the environment, going through similar stages of trial and error. Manipulation of the scene in the early stages was largely a matter of trial and error, but this has increasingly given way to more purposive procedures arising from observation of cause and effect and calculation of the results of actions taken. Hit and miss methods have given way to the scientific approach. It is not only man's early attempts to control the environment that can be described as hit and miss, the whole of unmanipulated ecology can be so described. Since the earth started, the development of its environments and of the life forms that go with them has been entirely fortuitous. The fact that a green and pleasant land came into being, and that the creatures living out their generations on it, without being aware of it probably derived enjoyment and pleasure from it, is beside the point. The point is that there was no creature on earth capable of mastering the laws which governed physical and biological processes, and in this sense the whole exercise was pointless, and existence was without purpose. Some would argue that this is how it should be, that life and man himself have come into being in the course of evolution; that man is merely a complex physico-chemical machine, with developed powers of widely distorting the machinery of nature; that in the course of time, the earth will run its course and disappear, and man with it. This is a pragmatic view, not shared by man's recent ancestors with their rather grotesque ideas of magic and religion, nor by a majority of the human race today. Scientists who argue in this way ignore that, however sophisticated, science only observes and measures what is already there. It can show the way to manipulate matter and energy to produce desired effects; it creates nothing and can explain nothing, except that which lies within the realm of a fore-ordained mathematical system.

Science consists of the observation of facts, that is of what happens in an ordered system; the analysis of those facts and the formation of hypotheses arising from them; and the testing of such hypotheses by further observation and mathematical analysis. Technology consists of the application of established facts to the construction of useful devices, most of which in some way or another will affect the environment. If these are the latest trends of ecology, they need not be written off either as damaging or pointless. Indeed, Bishop Berkeley postulated as long ago as 1710 that nothing exists unless there is an 'observer' of it.[1]   J. W. Dunne developed the same idea in relation to human consciousness, and the past and future, in *An Experiment with Time* and *The Serial Universe*.[2] If, therefore, one believes that there is any object in existence and consciousness, one must believe also that, sooner or later, an observer of the 'system' must appear.

I say 'of the system' advisedly, because what we can observe is this one mathematical system which rules our lives and destinies and which our senses, aided by the technological aids we can construct, are capable of observing. It is unthinkable, for instance, that miracles are 'paraphysical'; phenomena cannot occur *within this system* which offend against its laws; if they offend against its laws, they are outside the system. In this sense, scientists and mathematicians are rightly sceptical of such phenomena, until they can find an explanation for them. However, 'nature abhors a vacuum', and both scientists and mathematicians appear obsessed with what happens to their mathematical laws when there is a void; at the other end of the scale, they worry about what happens to them at infinity. Religion also worries about the infinite, probably with more reason. This is very odd, because known physical laws show that neither total vacuum nor infinity can exist within our system. The lower end of our scale is represented by absolute zero temperature ($0°$Kelvin) at which particulate matter loses energy; the upper end is represented by the speed of light at which matter loses its physical properties. Multiply the wavelength of energy waves by their frequency, and you inevitably get the speed of light. What then happens below or above these two points? There may plainly be other mathematical systems with different rules which science is unable to observe or measure. We can only say that they do not exist because we cannot observe them. But can we not observe them? An occult faculty of the human mind could possibly permit the observation of events in a different time scale and account for mystic experiences which it is impossible to ignore but which we cannot explain. Meanwhile, science has only probed a small proportion of the wave range in the mathematical scale, so that much remains to be discovered within our own mathematical milieu.

It may be asked — what has this to do with ecology, earth's environments and the future of the human race? It has, in fact, everything to do with it. Just as there exists a dividing line between the world of politics and science, so there is a dividing line between science and philosophy, including religion. It is important to the future of the world that scientists should realise and admit what many deny, that there could be events which affect us, occurring outside the mathematical system which rules our universe. The possibility of seemingly impossible events could then at least be recognised. If we aspire to control and manipulate our environment, the recognition of such possibilities becomes important. Hesitant steps are being taken to explore so-called extra-sensory perception (ESP), though this is confined to discovering whether it exists or not and not to ascertaining its mechanism. If there is a mathematical plan which requires an observer, then the observer too requires an observer, as was postulated by J.W. Dunne.

Science, to become a more effective force in environment planning, must shed this isolationism. It cannot spread its doctrines if it wraps their meanings in a mystical jargon, unintelligible even to fellow scientists in related disciplines. Incursions of physicists into biology have resulted in unravelling the mysteries of the genetic code. In other ways too, there is a trend away from isolationism amongst scientists, and one hopes to see this trend extend beyond the sciences. Are not finance and economics also subjects susceptible of scientific treatment? Are they not related to the behavioural sciences, and are not the behavioural sciences important to planning of human ecology? Should science shrug off mysticism and religion? In all these ways, one may hope that science in future may better help society to overcome its difficulties. Scientists do not communicate satisfactorily with their fellow men, and their fellow men are dubious of what they have to communicate.

How the present mundane turmoil will end, nobody can foretell. Ecological laws would suggest that we are at the present coming to the apex of a population explosion, following an era of great prosperity. We see around us the signs of the uneasiness and 'stress' that would be expected before the population crash. Cycles of explosion and crash, prosperity followed by decline both of living and moral standards are all too evident throughout human history, though the cycle is not one of years as with lemmings, but hundreds of years. This need not be restated, since it becomes evident even from a rudimentary study of human history and has been operative since Neolithic times. The question arises as to whether nature will take its course, or whether man, aided by his new-found technology, can avert it. Decline would, of itself, cure many of the problems of pollution, inflation, aggressiveness, and feelings of uncertainty, inadequacy and instability. To avoid decline, man must by his own efforts achieve a state of equilibrium with his environment, embracing not only advanced Western communities, but also the more backward communities of the world. The world of man is today a single ecosystem and population must be matched to resources.

In human affairs, population equilibrium is no longer achieved by matching death rate with birth rate; birth rate must be matched with a declining death rate. If birth rate is not matched to the needs of population replacement, then we shall lose our comfortable lives and our affluence, and in some way, which will be unpleasant, nature will do the job for us. This is so well known, that it might be thought superfluous to state it. Yet limitation of births by artificial means is still repugnant to some people on religious grounds, and politicians are surprisingly hesitant to take those steps that are open to them.

Both religious and secular laws admit freedom of choice to the individual over sexual relationships. However, emotive forces are evoked whenever proposals are put forward to make contraceptive advice and devices available to the unmarried. This is illogical, and in any case, once women were emancipated it was, or should have been, obvious that they would claim the same sexual freedom enjoyed by the male, if risks of unwanted pregnancy could be avoided. This development has not led – as it might have done – to a great increase of promiscuity. Most young people establish stable relationships with a partner as an experimental prelude to an association which in most cases has considerable durability. The husband is no longer regarded as the first bedfellow, but the last.

The ecologist may spare himself a quiet chuckle, because such was entirely predictable under ecological laws! 'Pair-bonding', as it is called by ethologists, is deeply ingrained not only in social custom, but in the very nature of the human race. Polygamy has, admittedly, been secondarily developed in some of the races of man, mostly as it would seem from economic causes. In most races from the least to the most advanced, one man and one woman live out their adult lives together sharing the good and the bad, and raising the family. Both make great sacrifices for the sake of the children, in order that they may be well equipped both physically and mentally in the life that lies ahead of them. This does not seem to be incompatible with a certain degree of promiscuity in some cases.

This is quite unlike what happens with our nearest relatives the anthropoid apes, except for the gibbons. Chimpanzees, in particular, are quite promiscuous; the female mates with a number of males when at oestrus, and there is no pair-bonding. The tendency to monogamy in man must have had survival value during the period when his ape-like ancestors were forced to leave the trees and adopt a terrestrial mode of life. The advantage must have lain in the preservation of the family unit, in the teaching of love and duty, and of useful skills to the young. The value of the family as a unit was never more clearly shown than today, when the unit is no longer so binding as formerly, and it is the children from broken families that provide most of our bad citizens and delinquents. Evidently also, sex and the performance of sex between partners came to be an important factor in cementing the pair bond between male and female. In no other primate is there the same marked difference between the sexes, a difference which leads to such an intense mutual desire for each other of one *particular* man and one *particular* woman, which adds up to what we call 'love' and leads to the stable relationship. How many times do the selected pair mate for each child born? Plainly, sex is as important socially in human affairs as for reproduction. This must have come about at least 100,000, if not a million, years ago.

It must be asked, when crystal-gazing into the ecological future, whether the pair-bonding system retains its value and will survive, or will it lose its value for urbanised societies and disappear? It is something which man and woman still plainly value, and most wish to retain. Nevertheless, promiscuity and broken marriages are on the increase, and these could indicate that an increasing proportion of urbanised society no longer values the pair-bond link, and would prefer a free-for-all. With family planning, fewer children are born, and over a shorter period, so that the family unit might come to have less value for their welfare.

It is fashionable amongst prophets of doom to paint horrifying pictures of the present-day breakdown of society and standards of behaviour. However, analysis of these trends hardly bears out these gloomy prognostications. A study of human history since the Christian era shows how much more tolerant and better off we are than ever before. Education and medicine are available to all. Every boy and every girl born into Western society has at least a reasonable chance of living his or her natural life span in health and contributing to human welfare. We do not kill the first born nor sacrifice our enemies or unfortunate virgins to the gods. Perhaps we have advanced too far in tolerance. If so, this is a revulsion against the wars and barbarities of the past. Some groups and individuals do take advantage of this tolerance and often punishment does not seem to fit the crime. What will be the effect on society as a whole of a failure to eliminate anti-social groups or persons? Society is at present seeking the answer. Crime and addictive habits are regarded as diseases requiring therapy rather than revenge. Obviously, the human animal is still badly adjusted to urban environments; ecology and evolution may have the last word in promoting greater adjustment over the generations. Adjustment of population levels, of behaviour patterns, and of general stability would be expected, if ecological laws continued to operate and man's environment continued to be sufficiently stable. As we saw in the last chapter, evolution still seems to operate on *Homo sapiens.*

The most frightening aspect of ecology is the total failure of world leaders, both national and international in political and economic spheres, and their inability to understand the simplest facts of science and technology. Political science must match physical science, if a stable environment for earth's inhabitants is to be planned. But, will a new type of planned environment remain stable even if it is achieved? Stability depends on two sets of factors: first, those that man can control; secondly, those beyond human control. Undoubtedly, man can adapt his environment to achieve equilibrium and stability. Whether he is temperamentally capable of doing so is another question and may be doubted. If he does not do so, he will either disappear into biological history, or enter a new dark age to re-emerge in a thousand years or so, and repeat the attempt. If successful, man faces a number of challenges in the technological sphere in the not-too-distant future. His comfortable life is attained by living on capital, by the use of fossil fuels laid down on earth in former biological ages. The greatest challenge to his technology is imminent, when he must find the means to tap alternative sources of power, sources that are income and not capital, and will, therefore, be lasting. The use of atomic energy presents many problems, and possibly the only lasting source of power will prove to be that of the sun trapped outside the earth's atmosphere, that power which originally gave birth to life itself. This possibility is, of itself, a justification for the exploration of space techniques.

Of the influences on human history of the future that lie outside the powers of man to manipulate, the most serious is that of climatic change. We do not know whether our present climate represents the end of the last ice age, or an intermission. If the latter, the glaciers will again descend from the north with a force and inevitability that no human effort can counter. Judging from glacial cycles, this could occur in 5,000 to 10,000 years from now. The northern hemisphere could be glaciated almost as far as the Mediterranean and the Gulf of Mexico. Floods would occur in the tropics and the Sahara would become well watered and fertile. 'The best laid plans of mice and men gang aft agley.'

As an epilogue to this work, it will be advisable to consider the earth as a single ecosystem of the human biosphere. If the earth is to be regarded as such, problems of population, food production, pollution, finance and economics are incapable of solution, unless mankind as a whole acts in unison, and unless world leaders of the necessary stature appear. At present, the outlook for such is gloomy. The peoples of the world are divided into warring factions. The Western civilisations are tensed, uneasy and stressed amidst unprecedented prosperity; 'iron-curtain' countries are bellicose; the rest of the world is undernourished and impoverished. How can the uniformity required by a single ecosystem come into existence under such conditions? Geneticists have been vocal on the subject of race differences and their responsibility for the existence of these conditions. Strangely, ecologists have been silent on the subject. So let us examine the question against an ecological background and in relation to two main topics: first, the problem of immigrant alien races; secondly, the differing abilities of the races.

Both zoology and ecology show that conflict arises when there is immigration of an allied but different race or species of animal into an established habitat. We have already studied the teachings of Alfred Russell Wallace of the effects on each other of races developed in isolation. It should not surprise us, therefore, when the same occurs

112

in human habitats. Whites are resented in black areas, and blacks in white areas. Jews are resented by their cousins, Semites, the Arabs. The trouble is less pronounced when the two races occupy different 'niches' in the habitat. To a certain extent, this happens with the Jews in Western societies, with their skills in trade and finance. However, their ability to accumulate wealth arouses resentment which at times boils over. Coloured peoples tend to occupy the niche of menial work, and they naturally resent their exclusion from more skilled jobs. However, if they emerge from this niche, they invade the niche of the white worker and are in their turn resented. This is in conformity with the ecological 'Principle of Competitive Exclusion', which states that two dissimilar races cannot occupy the same niche in the habitat. Those who ask for multi-racial states, in which the different races share equally, ask for the suspension of this natural law. Man has defied ecological laws before, and perhaps he can do so again. However, I cannot see Black Africa accepting while immigrants, or Indians or Chinese for that matter, on an equal basis with its own nationals, even if some semblance of integration were achieved in predominantly white countries.

If different niches cannot be found for competing races, ecological laws demand one of three other solutions:

1. The immigrant population will replace the existing one, of which many instances can be cited in human history;

2. The existing population will drive out the immigrant, as eventually the Spaniards did to the Moors;

3. The immigrant population will be assimilated.

The most recent example of assimilation is that of the United States in which different white races have been welded into a new nation. In countries such as Brazil, the coloured peoples have also been assimilated with a high degree of success. Britain, too, has assimilated waves of refugees and trading communities, who after a generation or two are more British than the British. With races such as the Jews, dedicated to preserving a national and religious identity, and coloured peoples, who have marked physical differences, assimilation becomes much more difficult. In the course of time, probably a long time, possibly they too might be assimilated, and this may be happening in the United States where white blood is much diluted by black. It has also been happening for many generations in parts of Africa, such as coastal Ghana, where black blood is much diluted by white. However, the problem is made more acute because of the widespread belief amongst whites that the black races are mentally inferior. This is the second problem which should be examined from the ecological viewpoint.

Scientists are often not particularly convincing when they stray from their chosen field. This remark applies especially to geneticists, when they stray into the realms of sociology. On the basis of purely sociological intelligence tests, some geneticists claim that black races are intellectually inferior to white in respect of IQ. Others maintain that the apparent difference is not genetical, but the result of social disadvantage and a poor background in childhood. It would appear from the evidence that both are partially right, but in any case from an ecological point of view the argument is completely sterile and argued on the wrong bases. It is further said that persons of any race with an IQ of under 100 should be given financial inducements to be sterilised. So, we are all to become Einsteins, and there will be no one to deliver the coal! This sort of argument is as dangerous as the proposal to alter genetic constitution by so-called 'genetic engineering'. Moreover, any such procedure would certainly not have the desired effect.

If we accept that man is a single species within the genus *Homo, H. sapiens,* and this we do, then all races share the same range of genes. It follows that differences of race, both mental and physical, are due not to gene *differences* but to a different distribution and combination of the same genes within the human 'polymorphic genetical system'. This is a very different thing. Man's 'gene pool' is enormously wide, and it is to this wide gene pool that he owes the many varied abilities which go to make his culture. His culture is dependent on a complex social structure, which in turn is dependent on the wide stretch of abilities which the wide gene pool makes possible. During the long and difficult centuries of the ice ages and after, evidently there was a genetical discrimination in the white races in favour of selecting and coupling those genes which make for higher intelligence. Undeniably, the white and other northern races have proved the more inventive and more capable of manipulating an unfriendly environment.

Faced with the challenge posed by Western civilisation and urbanisation, one would expect that a recoupling of genes would rapidly take effect in those races which had not experienced this particular type of challenge previously. No fundamental genetic change, as by mutation, is involved, merely a reshuffling of existing genes. Anybody who breeds animals knows how quickly new races or varieties can be established. IQ tests, therefore, whatever they show or prove, are totally irrelevant to the problem, though it is silly to suggest, as some do, that the problem does not exist. The ecological response in backward races to a changed environment should be adaptation to it, within a comparatively small number of generations.

Ecological adaptation is not altogether painless. While the polymorphic genetical system will work for a greater spread of abilities and increased powers of leadership and performance at one end of the scale, it also leaves a trail of inadequates at the other. This phase of the operation is evidently particularly active at the present time amongst urbanised Western communities. As we have already seen, these unfortunates tend to be 'genetically unfit' and so in the long run discarded. There might also come a time in human history when physical endurance was more important to survival than high intelligence. In this event, IQ would recede and the robuster individuals would be the survivors. It is this flexibility of the polymorphic genetical system which is of such importance. A change of gene structure by mutation would take thousands, even tens of thousands, of years to alter race characteristics, against a few generations by gene reshuffling.

Every gene has one factor from the father and one from the mother. These may be dissimilar, and each gene is recombined at each generation. Moreover, many characters, particularly perhaps intelligence, are determined by more than one gene. Furthermore, there are other 'modifier' genes, which modify genes' effects. So to talk of selecting human beings for intelligence is utter nonsense; only the challenge of the environment can do this. Wipe out a country's intelligentsia, as has happened in France and Russia; within a generation or two, equally gifted leaders have arisen. At any rate, how profoundly foolish it would be to select man for one or two characters, when his social structure demands such a wide spread of abilities and his survival is dependent on the ability of his genetical system to adapt the physical characters of the race quickly to changed environmental circumstances. In general, white races have become adapted to life in temperate climates. They are not readily adapted to life in tropical regions; while the first generation performs effectively, the second and succeeding generations decline in vigour. Conversely, the black races have become adapted to life in hot countries, and in the absence of refreshment by newer immigration it is doubtful whether their numbers would be maintained in temperate climates. During the sixteenth and seventeenth centuries, large numbers of Africans had been brought to Britain; by 1939, they had disappeared. Possibly they were absorbed into the general population; possibly they declined because of greater susceptibility to infectious diseases, such as tuberculosis and pneumonia, or to nutritional diseases such as rickets. During my travels in the United States, I have found the negro people more contented and more in tune with the environment in the south than in the north, in spite of supposed greater intolerance there.

The lessons of ecology would appear to suggest that the different races will perform best in the areas which have produced them. In such areas, I do not doubt that the backward peoples, in response to the change of the environment, will quickly become equipped with the same range of talents that has appeared so quickly in Western and Eastern cultures. The signs are already to be seen in African universities.

It is, in any case, profoundly to be hoped that this forecast is correct. The world, as a single ecosystem of the human biosphere, cannot afford two standards of culture, the rich nations and the poor nations, any more than Victorian England could afford Disraeli's two nations. Ecological disequilibrium means instability. It seems to me that the solution of environmental problems is dependent, more than anything else, on the ability of the races of man to develop their maximum potential within their own niches and in harmony with each other, so that world problems can be tackled jointly.

NOTES

1 Bishop G. Berkeley, *Principles of Human Knowledge*, 1710
2 J.W. Dunne, *An Experiment with Time*, Faber, London, 5th edn., 1939, and *The Serial Universe*, Faber, London, 1934

Further Reading

Carthy, J.D. and Ebling, F.J. (eds), *The Natural History of Aggression*, Inst. of Biol. Symposium, Academic Press, London & New York, 1964
Montagu, A., *Human Heredity*, Signet Science Library, New York, 1963
Parlin, C., *The Relations between the Sciences*, University Press, Cambridge, 1968
Wolfgang, W., *The Sexual Code. The Social Behaviour of Animals and Men*, Doubleday, New York, 1972

INDEX

119

DEFORMATION AND STRENGTH
OF MATERIALS

DEFORMATION AND
STRENGTH
OF MATERIALS

P. FELTHAM
D.Sc., F.Inst.P.
Brunel University, London W.3

Springer Science+Business Media, LLC
1966

Suggested U.D.C. number : 539:3/·4

Originally published by Butterworths in 1966.
Softcover reprint of the hardcove 1st edition 1966

ISBN 978-1-4899-5849-5   ISBN 978-1-4899-5847-1 (eBook)
DOI 10.1007/978-1-4899-5847-1

PREFACE

ONE of the effects of the unprecedented advance in the synthesis and use of new materials in the last few decades, but particularly since the end of the Second World War, has been the widespread introduction of 'materials science' into curricula of university courses. The momentum persists, and from the point of view of research and applications a vast array of properties of alloys, ceramics, semiconductors, polymers, fluids, cellular and other materials remains in the focus of interest.

How to distil from this embarrassing wealth of matter the essential features necessary for an adequate interpretation of their properties, how in fact to introduce the theoretical foundations of materials science into university curricula in an effective yet concise manner has therefore been a topical problem for some time.

In the present textbook, which is based largely on lectures to science and technology undergraduates, I have attempted the task of dealing systematically with the aspects of materials science which, particularly in the light of my experience in industrial research, appeared to me to be essential to an understanding of the fundamental fabric of principles. The boundaries of the subject are not well defined; a study of materials must have its roots in the firm soil of the classical sciences. An elementary knowledge of mathematics has therefore been assumed. Topics well covered in standard texts on physical chemistry, as well as problems involving electron transport, which are ably dealt with in many books on 'solid state' have not been included. The emphasis has been on the mechanical properties of materials, including elasticity, visco-elastic behaviour, damping capacity, the strength of real crystals, dislocation theory, fracture, fatigue, and the behaviour of non-Newtonian fluids.

Wherever possible I have tried to show the connection between different forms of behaviour, and to deduce quantitative relations in the simplest possible way. Although the theoretical foundations of macroscopic elasticity and plasticity are well established, the micromechanisms of plastic flow and fracture, for example, have not as yet been fully elucidated in many materials, and they may in fact be subjects of keen controversy. This state of affairs is responsible for the tentative approach which the reader will occasionally find in the book, for example in the treatment of work-hardening.

The book is aimed at a large circle of readers interested in the theoretical foundations of the subject, and it should prove particularly useful to students in the science, technology and medical faculties of universities. The emphasis has therefore been on the clear and concise explanation of the essential, basic, concepts of the mechanical properties of materials, and not on the detailed survey of the properties of the great variety solids and liquids encountered in practice. Its function is considered to be an exposition which should enable the serious reader to learn the language of the subject. Thus armed he can then search with profit in papers and specialised handbooks for guidance on more specific problems.

London   P.F.

CONTENTS

vii

ELASTICITY

1.1 Introductory

Materials which can be subjected to significant deformations, but which regain their original shape when the constraints are removed, are solids. This complete recovery of form, observed at least up to certain levels of the stress, characterising the elastic range, distinguishes them from liquids, which will suffer irreversible changes of shape in any general deformation, however small.

Experience shows that within the elastic range the relation between the deformations and the applied stresses is linear; the relationship is known as Hooke's Law. Beyond the elastic range the material acquires a permanent set; a plastic deformation is said to remain when the forces are removed.

Figure 1.1 shows a stress-strain curve typical of many materials. It can be obtained by applying a steadily increasing tensile stress σ to a rod or wire, and plotting the tensile strain ϵ in the course of deformation. The elastic range terminates at σ_e, while gross plastic yielding occurs above the yield stress σ_y. If at a certain point P the stress is removed, and the test is subsequently continued, the specimen would behave elastically as indicated by the broken line AP, and would then extend from P onward as if no interruption had taken place. The total strain can be seen to consist of the plastic component, OA, and the elastic contribution AB. Young's modulus is given by tan α, which represents the slope of the curve in the elastic range.

The stress at the point P is termed the flow stress at the corresponding strain, to distinguish it from the yield stress which is measured at the initial transition to the plastic state. As this transition may not be sufficiently abrupt to enable one to ascribe a fairly exact value to σ_y a proof stress is generally specified instead in engineering practice, particularly with metals. This represents the value of the flow stress at some small total tensile strain, such as $0 \cdot 1$ or $0 \cdot 2$ per cent, and provides a practical measure of the load-bearing capacity of the material. In general it is only

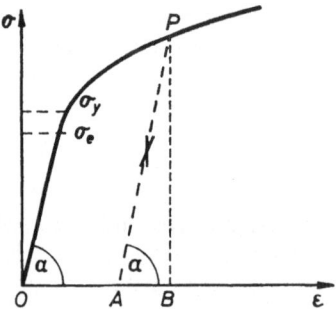

Figure 1.1. Stress-strain curve of a ductile material

slightly higher than σ_y. The smallness of these reference strains gives an indication of the limited extent of the elastic range in crystalline materials, which contrasts sharply, for example, with rubber-like solids.

Structural changes induced by the deformation result in work-hardening, apparent in the need to apply increasingly higher stresses to strain the material as the tensile test proceeds. These changes eventually lead to fracture. If the total strain ϵ_f attained at the instant of fracture is large compared with the elastic component the material is said to be ductile, otherwise it is regarded as brittle. The strain at fracture is in fact a convenient practical index of ductility.

2

At temperatures relatively high with respect to the melting point (T_m), say above about $0 \cdot 3 \, T_m$ (°K), the work-hardened structure generally becomes unstable. The material softens, anneals, in the course of time, and hence also during the test, and the shape of the stress-strain curve becomes appreciably strain-rate dependent. In an unloading-loading cycle, as described above, the initial and final points on the curve may then no longer coincide, and altogether the practical usefulness of work-hardening curves then becomes limited.

Elastic deformation does not only contribute to the overall change of shape of materials subjected to stresses, but it is also a prerequisite for the occurrence of permanent changes of shape in solids. The question of the quantitative representation of states of stress and strain in elastic materials is therefore of basic importance here.

1.2 Uniaxial Stress

A body which is deformed by the application of a force is automatically subjected to an equal and opposite reaction and, in view of the continuity of the material, a state of stress will exist at all points. In analysing this stress it is sufficient to consider the applied forces explicitly, for whatever generalisations are deduced for them applies equally to the reactions.

Now, a force acting on a small area of a solid can be resolved into two orthogonal components, one acting in the surface, the other in a direction at right-angles to it. The first component, together with its reaction, will give rise to a shear stress, the latter to a tensile stress. Although the force is a vector, the system of stresses induced by it is not. This may be seen quite readily by considering that an element of the surface of a body, sufficiently small to be regarded plane, would be specified in a rectangular coordinate system by the direction of its normal which, like the force acting on it, is a vector quantity. The stress,

3

defined as the force per unit area, is therefore represented by the ratio of two vectors, which is not a vector itself, but is known as a tensor. Before considering this question further we shall examine some simple states of stress, beginning with a cylindrical rod of circular cross-section subjected to uniaxial tension by a force F, as shown in *Figure 1.2.*

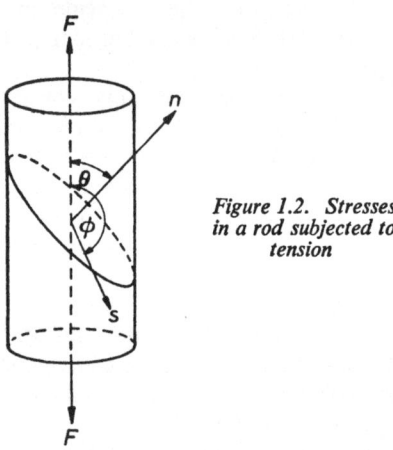

Figure 1.2. Stresses in a rod subjected to tension

The force acting at right-angles to a plane having normal n is $F \cos \theta$ and the area of the ellipse on which it is acting is $A/\cos \theta$, where A is the area of cross-section of the rod. The tensile stress on planes making an angle θ with the axis is therefore $(F/A) \cos^2 \theta$, with a maximum $\sigma = F/A$ on surfaces perpendicular to the rod axis. The tensile stresses may therefore be written as

$$\sigma(\theta) = \tfrac{1}{2}\sigma(1+\cos 2\theta) \qquad (1.1)$$

Similarly, by considering the component $F \sin \theta$ acting on the plane of the ellipse in the direction of its major axis, the shear stress $\tau(\theta)$ is found to be $(F/A) \sin \theta \cos \theta$, or

$$\tau(\theta) = \tfrac{1}{2}\sigma \sin 2\theta \qquad (1.2)$$

4

A geometrical representation of the stresses as function of the angle θ, known as Mohr's circle, is shown in *Figure 1.3*.

It may readily be verified that the co-ordinates of a point on a circle of radius $\frac{1}{2}\sigma$, with centre on the τ-axis and touching the σ-axis are in fact given by equations 1.1 and 1.2 if the radius vector drawn from the centre of the circle to the point P makes an angle 2θ with the shear stress axis. In view of the simplicity of equations 1.1 and 1.2 Mohr's circle is not here of material assistance in the

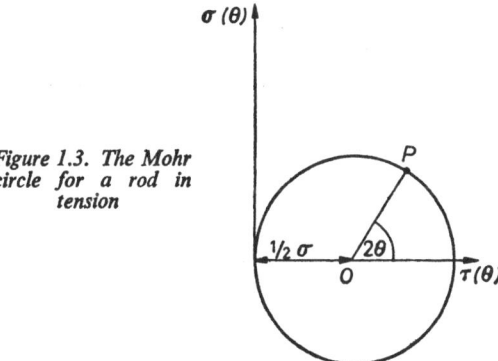

Figure 1.3. The Mohr circle for a rod in tension

determination of stresses, but geometrical methods based on the extension of this principle are frequently used in analyses of more complex systems of stress and strain, and in problems of fracture.

Equation 1.2 implies that the maximum shear stress is equal to $\frac{1}{2}\sigma$, and that it occurs on planes inclined at 45° to the tensile axis. This result has important implications in connection with the mode of plastic deformation of ductile materials, as will be seen later.

1.3 Biaxial Stress

A more complex state of stress results from the simultaneous action of two orthogonal tensile stresses. Close approximations to such states of stress may occur, for example, in stretched membranes or in the shells of thin-walled pressure vessels.

Our considerations will apply irrespective of whether one or both of the stresses are negative, i.e. compressive; results for any specific case are obtained simply by the use of the appropriate signs for the stresses. We shall denote these here by σ_1 and σ_2, and take $\sigma_1 > 0 > \sigma_2$. The diagrams in *Figure 1.4* show that this system can be split into two components, one of which is a uniaxial tension, and the

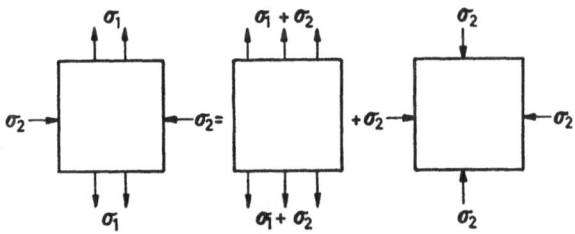

Figure 1.4. Resolution of a biaxial stress

other a two-sided compression. The first will give rise to shear and tensile stresses analogous to those given by equations 1.1 and 1.2, with $\frac{1}{2}(\sigma_1+\sigma_2)$ substituted for $\frac{1}{2}\sigma$, and, of course, to corresponding changes of shape, in this case to an axial elongation. The second system will not induce any changes in shape of the membrane or foil, except in the thickness (neglected in this case), but with compressive stresses as in *Figure 1.4*, a square lamella would become smaller in size, yet at the same time it would remain square. The system could again be represented by a Mohr circle. The radius would be $\frac{1}{2}(\sigma_1+\sigma_2)$, and the centre would lie on the τ-axis at a distance $\frac{1}{2}(\sigma_1+\sigma_2) - \sigma_2$ from the origin.

The system could be split up in other ways, for example by isolating σ_1 rather than σ_2. This arbitrariness disappears

6

if the system is considered to be a special case of a triaxial stress with $\sigma_3 = 0$, as will become apparent below.

1.4 Deviatoric and Isotropic Stresses

In the three-dimensional stress system we shall consider three mutually perpendicular stresses σ_1, σ_2 and σ_3 acting at right-angles to the faces of a small cube of material. We now wish to separate the component which gives rise to changes of shape, termed the deviator, from the isotropic, or hydrostatic, stress responsible only for changes in volume.

Now, writing σ_h for the hydrostatic stress, one may represent the separation formally by

$$(\sigma_1, \sigma_2, \sigma_3) = (\sigma_1 - \sigma_h, \sigma_2 - \sigma_h, \sigma_3 - \sigma_h) + (\sigma_h, \sigma_h, \sigma_h) \quad (1.3)$$

It is not obvious what combination of the three stresses is required to give σ_h, so we shall put

$$\sigma_h = a\sigma_1 + b\sigma_2 + c\sigma_3 \quad (1.4)$$

where a, b and c are yet undetermined constants. Since however none of the three stresses is in any way preferred, a, b and c must be equal, so that it remains to determine the magnitude of a. To this end we consider a specific case in which all three stresses are very nearly equal so that the stress system consists almost entirely of the hydrostatic component. The deviator is then equal to zero, and we must therefore also have

$$(\sigma_1 - \sigma_h) + (\sigma_2 - \sigma_h) + (\sigma_3 - \sigma_h) \approx 0$$

This relation is compatible with equation 1.4 only if

$$\sigma_h = \tfrac{1}{3}(\sigma_1 + \sigma_2 + \sigma_3) \quad (1.5)$$

which is the required result.

It is clear from equations 1.3 and 1.5 that the deviator cannot now be represented by a uniaxial tensile stress, as in *Figure 1.4* ; all three stresses have to be taken into account.

The decomposition of the stress system is of considerable practical importance. It enables one to find the stresses responsible for changes of shape, and to study their effects separately from those introducing isotropic dilatations or compressions. Metals and most inorganic crystals, for example, may for many purposes be regarded as incompressible, and the contribution of the isotropic stress to the total deformation may often be neglected.

1.5 Stresses at a Point in a Body

It may seem that the stress systems so far considered in which, firstly, the stresses were mutually perpendicular and, secondly, shear stresses were not specifically introduced, are not sufficiently general to be of theoretical interest. However, we shall show that the state of stress in, at least, a small volume around any point of an elastically deformed body may be represented by three mutually perpendicular so-called principal stresses, provided the axes of reference are suitably oriented in the body. The required orientation is obvious in most cases occurring in practice, as we shall illustrate by way of an example below; the principal stresses therefore provide a full description of the state of stress over a volume around the chosen point within which the stresses may be considered to be uniform, without abrupt change.

The state of stress close to some point O in the deformed elastic material could be ascertained by a hypothetical experiment in the following way. A set of three axes is imagined in the body, with O as origin. A small tetrahedron is then cut out as indicated in *Figure 1.5*, the base plane ABC having some arbitrary but known orientation given by the angles α, β and γ which the normal makes with the x, y and z axes respectively. The tetrahedron is then imagined to be removed outside the body. As the initial constraints are then no longer acting, the tetrahedron will distort, and tensile and shear stresses must be applied to it faces to restore the previous shape. If these are measured, then the state of stress at the point O is in principle determined. By considering the equilibrium of forces acting on the tetrahedron we shall show that in general three tensile and three shear stresses are required to specify the stress at a point of an elastically isotropic body.

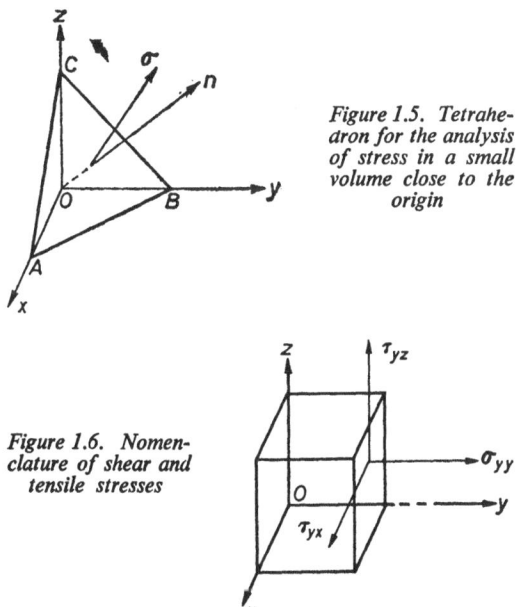

Figure 1.5. Tetrahedron for the analysis of stress in a small volume close to the origin

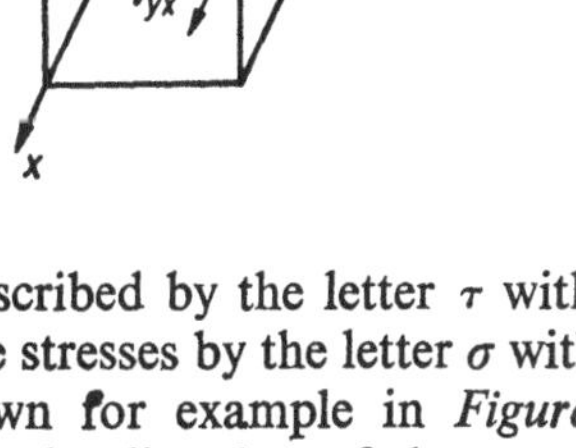

Figure 1.6. Nomenclature of shear and tensile stresses

Shear stresses will be described by the letter τ with two dissimilar subscripts, tensile stresses by the letter σ with two similar subscripts, as shown for example in *Figure 1.6*. The first subscript specifies the direction of the normal of the plane on, or in, which the stress is acting; for all stresses shown in the figure this is a plane parallel to $y = 0$. The second subscript indicates the direction of the force which gives rise to the stress. This direction is along the normal of the plane in the case of tensile stresses, and both letters of the subscript are then the same. Two different subscripts indicate a shear stress. The nature of the stress is therefore apparent from the subscripts, and the use of two different letters, namely τ and σ, is not essential, although it will be employed here for convenience. A normal stress is tensile or compressive, i.e. negative, depending on whether the two arrows representing it both point away or towards the element on which they act. The convention relating to the sign of shear stresses is as follows.

The couple representing the stress, for example τ_{yz} in *Figure 1.6*, is assumed to act about the origin. Then, if the arrow on the positive side of the origin, in this case at $y > 0$, points along the positive direction of the co-ordinate, i.e. z, the stress is positive; with arrows reversed it is negative.

Now, referring to *Figure 1.5*, we note that the angles which the plane ABC makes with the planes $x = 0$, $y = 0$ and $z = 0$ are α, β and γ respectively, and that consequently the areas of the triangles OBC, OCA and OAB are equal to $\Delta \cos \alpha$, $\Delta \cos \beta$ and $\Delta \cos \gamma$, where Δ is the area of the triangle ABC. Since the forces acting on the tetrahedron are in equilibrium their resolved components in any direction must also be in equilibrium. In particular, if σ_{nx} is the component in the x-direction of the stresses acting on the triangle ABC, then

$$\Delta \sigma_{nx} = \sigma_{xx} \Delta \cos \alpha + \tau_{yx} \Delta \cos \beta + \tau_{zx} \Delta \cos \gamma$$

or

$$\sigma_{nx} = \sigma_{xx} \cos \alpha + \tau_{yx} \cos \beta + \tau_{zx} \cos \gamma$$

and similarly

$$\sigma_{ny} = \tau_{xy} \cos \alpha + \sigma_{yy} \cos \beta + \tau_{zy} \cos \gamma$$

and

$$\sigma_{nz} = \tau_{xz} \cos \alpha + \tau_{yz} \cos \beta + \sigma_{zz} \cos \gamma$$

$$(1.6)$$

We now wish to discover whether we could find a plane,

like *ABC* above, on which no shear stresses act, the stresses on *ABC* being purely tensile or compressive. Let us assume that the plane *ABC* in *Figure 1.5* is in fact such a 'principal' plane. The stresses σ_{nx}, σ_{ny} and σ_{nz} are then equal to $\sigma_p \cos \alpha$, $\sigma_p \cos \beta$ and $\sigma_p \cos \gamma$ respectively, where σ_p is the principal stress acting on the triangle *ABC*. On substituting these values into equation 1.6 one obtains

$$\left. \begin{array}{l} (\sigma_{xx} - \sigma_p) \cos \alpha + \tau_{yx} \cos \beta + \tau_{zx} \cos \gamma = 0 \\ \tau_{xy} \cos \alpha + (\sigma_{yy} - \sigma_p) \cos \beta + \tau_{zy} \cos \gamma = 0 \\ \tau_{xz} \cos \alpha + \tau_{yz} \cos \beta + (\sigma_{zz} - \sigma_p) \cos \gamma = 0 \end{array} \right\} \quad (1.7)$$

which can be considered as three equations in the unknowns $\cos \alpha$, $\cos \beta$ and $\cos \gamma$, now determining the orientation of the principal plane with respect to the chosen reference axes. As is well known they will yield solutions, in this case real values of σ_p, only if the determinant

$$\begin{vmatrix} \sigma_{xx} - \sigma_p & \tau_{yx} & \tau_{zx} \\ \tau_{xy} & \sigma_{yy} - \sigma_p & \tau_{zy} \\ \tau_{xz} & \tau_{yz} & \sigma_{zz} - \sigma_p \end{vmatrix} = 0 \quad (1.8)$$

On expanding this determinant one obtains a cubic equation in σ_p, yielding three real values σ_1, σ_2 and σ_3, which are the principal stresses at the point O. It may readily be shown that they are mutually perpendicular. Hence, from a knowledge of the stresses acting in the planes of an arbitrarily chosen co-ordinate system at O the magnitudes of the principal stresses can be determined. If these are substituted in turn into equations 1.7 three sets of direction cosines are obtained. They determine the directions of the principal stresses with respect to the chosen reference system. The state of stress at a point is therefore fully described by six independent data, namely either the magnitude of the three principal stresses and the location of the principal axes at O (which necessitates specifying three angles) or by the shear and tensile stresses on the orthogonal faces of the tetrahedron (*Figure 1.5*). Although

11

it may appear from equation 1.8 that nine stress components have to be known to specify the stress, this is not in fact the case, for $\tau_{xy} = \tau_{yx}$, $\tau_{yz} = \tau_{zy}$ and $\tau_{zx} = \tau_{xz}$. This is clear if one considers that the cube in *Figure 1.6* would rotate counterclockwise about an axis through its centre parallel to the x-axis under the action of the shear stress τ_{yz} unless it is balanced by an opposing couple due to a shear stress τ_{zy}, with $\tau_{yz} = \tau_{zy}$.

Equation 1.8, but with σ_p omitted, tabulates the stresses at a point in an isotropic elastic material, and is referred to as the stress tensor; as we have seen, only six of the stresses are independent.

In practice the directions of the principal stresses are often easily identified, as we shall show by way of example below. One then obtains the simple representation of a triaxial stress system considered in section 1.4. By analogy with the uniaxial stress system (equations 1.1 and 1.2) the greatest shear stresses now act on planes inclined 45° to the pairs of stresses (σ_2, σ_3), (σ_3, σ_1) and (σ_1, σ_2) respectively; their absolute values

$$\left.\begin{array}{l} \tau_1 = \tfrac{1}{2} \left| \sigma_2 - \sigma_3 \right| \\ \tau_2 = \tfrac{1}{2} \left| \sigma_3 - \sigma_1 \right| \\ \tau_3 = \tfrac{1}{2} \left| \sigma_1 - \sigma_2 \right| \end{array}\right\} \tag{1.9}$$

are known as the principal shear stresses.

1.6 Invariants of the Stress Tensor

The cubic equation in σ_p, obtained on expanding the determinant in equation 1.8, may be written formally

$$\sigma_p^3 - I_1 \sigma_p^2 - I_2 \sigma_p - I_3 = 0 \tag{1.10}$$

where

$$I_1 = \sigma_{xx} + \sigma_{yy} + \sigma_{zz} \tag{1.11}$$

and I_2 and I_3 are homogeneous expressions of second and third degrees in the stress components, respectively, as is readily verified. Now, the principal stresses at the point O, which are the roots of equation 1.10, clearly do not depend upon the initial choice of the co-ordinate system at O. Consequently the coefficients in equation 1.10 must also be independent of the choice of axes. It follows that if a differently oriented set of axes had been chosen, for example the principal axes, then one should have

$$\sigma_{xx} + \sigma_{yy} + \sigma_{zz} = \sigma_1 + \sigma_2 + \sigma_3 \qquad (1.12)$$

so that the sum of three mutually perpendicular tensile stresses in any Cartesian co-ordinate system centred on O is equal to I_1 – a constant. Similar conclusions may be drawn for I_2 and I_3. I_1, I_2 and I_3 are known as the first, second and third invariants of the stress tensor of an isotropic elastic material. Reference to equation 1.5 shows that

$$I_1 = 3\sigma_h \qquad (1.13)$$

1.7 Elastic Constants and Moduli

Before considering the relation between stress and strain it is necessary to specify the description of strains. As in the case of stresses two types of strain are distinguished, namely shear and tensile.

When dealing with small elastic deformations a tensile strain in any given direction in the body is given by the elongation δl in that direction divided by the initial length l_0. Thus, for a rod with axis along the x-co-ordinate an elongation will give rise to a tensile strain $\epsilon_{xx} = \delta l / l_0$ and, due to the Poisson contraction, to compressive normal strains at right-angles to the axis; thus ϵ_{yy} and ϵ_{zz} will be negative. Referring to the cube of elastic material shown in elevation in *Figure 1.7a*, the shear strain γ_{zy} due to the shear stress τ_{zy} is generally defined in engineering practice by the tangent of the angle AOA', where AA' is the displacement of the upper cube surface. As the strains considered are small the tangent may be replaced by the angle, expressed in radians. The deformation shown in *Figure 1.7a* is characterised by the fact that any plane in the cube parallel

13

to $z = 0$ is displaced rigidly along a straight line, in this case the y-direction, and the separation of any pair of such 'shearing planes' remains constant in the course of deformation; the cube does not extend or contract along the z-axis. Such a shear is referred to as 'simple'. However, as can be seen from *Figure 1.7b* the same change of shape could also be introduced in a symmetric manner by displacing OA and OC in opposite senses such that the angles AOA'' and COC'' are each equal to $\frac{1}{2}\gamma_{zy}$.

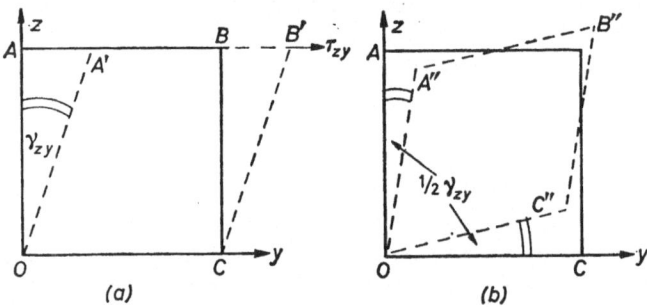

Figure 1.7. (a) *The engineering shear strain;* (b) *rational shear strain*

The angle AOA'' represents the rational shear strain, generally used in the mathematics of elasticity. The deformation which it defines (*Figure 1.7b*) is a 'pure' shear. It describes the change of shape of the elastic cube; the simple shear can be obtained by an additional rotation of the deformed cube through an angle $\frac{1}{2}\gamma_{zy}$ in the clockwise sense about the origin, as is apparent from *Figure 1.7*. A rotation not affecting the shape of the body is therefore implied by the engineering definition of strain, which refers to *Figure 1.7a*.

The principal axes in *Figure 1.7b* lie in the directions $A''C''$, OB'' and the x-axis, and do not change their directions in the course of the deformation. This constancy of the

14

directions of the principal axes is the main feature differentiating a ' pure ' from a ' simple ' shear. It is readily verified that the principal stresses are

$$(\sigma_1, \sigma_2, \sigma_3) = (- \tau_{zy}, \tau_{zy}, 0) \qquad (1.14)$$

with σ_1 along $A''C''$ and σ_2 along OB''. In this system τ_{zy} is then the principal shear stress τ_3 as defined by equation 1.9. As in the case of stress it may again be shown that six independent strain components are necessary, in general, to specify the state of strain at a point in an elastic, homogeneous, material; the relations $\gamma_{xy} = \gamma_{yx}$, $\gamma_{yz} = \gamma_{zy}$ and $\gamma_{zx} = \gamma_{xz}$ again apply. The strain tensor takes the form

$$\begin{vmatrix} \epsilon_{xx} & \tfrac{1}{2}\gamma_{yx} & \tfrac{1}{2}\gamma_{zx} \\ \tfrac{1}{2}\gamma_{xy} & \epsilon_{yy} & \tfrac{1}{2}\gamma_{zy} \\ \tfrac{1}{2}\gamma_{xz} & \tfrac{1}{2}\gamma_{yz} & \epsilon_{zz} \end{vmatrix}$$

the values of the rational shear strains being used.

In its most general form Hooke's Law may be expressed by representing the six components of the stress at a point by linear combinations of all the strains. One then obtains

$$\sigma_{xx} = c_{11}\epsilon_{xx} + c_{12}\epsilon_{yy} + c_{13}\epsilon_{zz} + c_{14}\gamma_{yz} + c_{15}\gamma_{zx} + c_{16}\gamma_{xy}$$

$$\sigma_{yy} = c_{21}\epsilon_{xx} + c_{22}\epsilon_{yy} + c_{23}\epsilon_{zz} + c_{24}\gamma_{yz} + c_{25}\gamma_{zx} + c_{26}\gamma_{xy}$$

$$\sigma_{zz} = c_{31}\epsilon_{xx} + c_{32}\epsilon_{yy} + c_{33}\epsilon_{zz} + c_{34}\gamma_{yz} + c_{35}\gamma_{zx} + c_{36}\gamma_{xy}$$

$$\tau_{yz} = c_{41}\epsilon_{xx} + c_{42}\epsilon_{yy} + c_{43}\epsilon_{zz} + c_{44}\gamma_{yz} + c_{45}\gamma_{zx} + c_{46}\gamma_{xy}$$

$$\tau_{zx} = c_{51}\epsilon_{xx} + c_{52}\epsilon_{yy} + c_{53}\epsilon_{zz} + c_{54}\gamma_{yz} + c_{55}\gamma_{zx} + c_{56}\gamma_{xy}$$

$$\tau_{xy} = c_{61}\epsilon_{xx} + c_{62}\epsilon_{yy} + c_{63}\epsilon_{zz} + c_{64}\gamma_{yz} + c_{65}\gamma_{zx} + c_{66}\gamma_{xy}$$

The 36 coefficients are known as elastic constants, and are tabulated for various crystalline materials for specific orientations of the stress co-ordinates; these are chosen to coincide, in general, with some of the symmetry axes of the crystal. Only 21 of the coefficients are independent, for it may be shown that $c_{mn} = c_{nm}$, irrespective of whether the material is isotropic or not. This equality is known as Onsager's Reciprocity Relation.

Clearly, it is also possible to express the six components

of strain in terms of the stresses. One then obtains 21 elastic moduli s_{mn}, m, $n = 1, 2, 3$. Each may be expressed in terms of elastic constants, and vice versa.

In the case of elastically isotropic materials, with which we shall be mainly concerned, we shall derive the relation involving the moduli in a representation referred to principal axes. It is of sufficient generality in this form for many applications, and complications due to the explicit introduction of shear stresses and strains are circumvented.

1.8 Hooke's Law referred to Principal Axes

The relation between the principal stresses and the corresponding principal strains ϵ_1, ϵ_2 and ϵ_3 are readily deduced from first principles. We note that the strain due to the application of the stress σ_1 to an isotropic, elastic, cube is σ_1/E, where E is Young's modulus. If now another principal stress is applied, say σ_2, the previous strain will be diminished by $v\sigma_2/E$ if σ_2 is positive, or increased by the same amount if σ_2 is a compressive stress. A similar effect would arise from the application of σ_3. Thus, taking Poisson's ratio v to be positive, one has

$$
\left.
\begin{aligned}
\epsilon_1 &= \frac{1}{E} \left[\sigma_1 - v \left(\sigma_2 + \sigma_3 \right) \right] \\[2mm]
\epsilon_2 &= \frac{1}{E} \left[\sigma_2 - v \left(\sigma_3 + \sigma_1 \right) \right] \\[2mm]
\epsilon_3 &= \frac{1}{E} \left[\sigma_3 - v \left(\sigma_1 + \sigma_2 \right) \right]
\end{aligned}
\right\}
\qquad (1.15)
$$

If this form of Hooke's Law is compared with the general representation, involving equations such as

$$\epsilon_{xx} = s_{11}\sigma_{xx} + s_{12}\sigma_{yy} + s_{13}\sigma_{zz} + s_{14}\tau_{yz} + s_{15}\tau_{zx} + s_{16}\tau_{xy}$$

and similar ones for the remaining five stress components, the simplification due, in particular to the elimination of the shear stresses, is clearly apparent. The assumption of isotropy has resulted in the reduction of the number of moduli to two, since, remembering that $s_{mn} = s_{nm}$,

$$s_{11} = s_{22} = s_{33} = 1/E \text{ and } s_{12} = s_{13} = s_{23} = -v/E$$

On rearranging equations 1.15 the stresses may be expressed in terms of the strains; the elastic constants are then found to be

$$c_{11} = c_{22} = c_{33} = E(1-v)/(1+v)(1-2v)$$

and

$$c_{12} = c_{13} = c_{23} = v E/(1+v)(1-2v)$$

The ratio $vE/(1+v)(1-2v)$ is sometimes denoted by λ, and

$$E/2(1+v) = G \tag{1.16}$$

Both λ and G are known as Lamé's constants; as we shall see later, G is the shear modulus.

The hydrostatic component of the strain, ϵ_h, accounts for changes in volume, but not in shape, and is therefore given by $\Delta V/V$, where ΔV is the change in volume of a cube of volume V under the action of the principal stresses σ_1, σ_2 and σ_3. If V is taken to be unity, ΔV is the hydrostatic strain. Now, taking $V = 1$:

$$1 + \Delta V = (1 + \epsilon_1)(1 + \epsilon_2)(1 + \epsilon_3)$$

On expanding the bracketed terms and neglecting terms of second and higher orders one obtains

$$\epsilon_h = \epsilon_1 + \epsilon_2 + \epsilon_3 \tag{1.17}$$

The relation

$$\sigma_h/\epsilon_h = K \tag{1.18}$$

with σ_h given by equation 1.5 defines the bulk modulus K, or its inverse, the compressibility. We note that for an incompressible material the volume strain must always be zero, so that then

$$\epsilon_1 + \epsilon_2 + \epsilon_3 = 0 \tag{1.19}$$

Since an equation analogous to 1.12 also holds for strains, one has for incompressible isotropic materials

$$\epsilon_{xx} + \epsilon_{yy} + \epsilon_{zz} = 0$$

independent of the choice of axes. In view of this relation the strain at a point of an incompressible elastic material is specified if five rather than six components of the strain tensor are known.

1.9 Energy Stored in Elastically Deformed Materials

The energy of elastic deformation may be regarded as potential energy, for it can be converted into kinetic energy, performing work as, for example, in a clockwork mechanism driven by a coiled metal spring. In the case of a wire or rod of cross-section A and length l_0 subjected to a tensile stress σ_1 the work stored is given by

$$w_1 = \int_{l_0}^{l} A \cdot \sigma \cdot \mathrm{d}l = A l_0 \int_0^{\epsilon_1} \sigma \, \mathrm{d}\epsilon$$

where $\epsilon_1 = \sigma_1/E$. Since Al_0 is the volume of the rod or wire, the energy stored per unit volume is given by

$$w_1 = E \int_0^{\epsilon_1} \epsilon \cdot \mathrm{d}\epsilon = \tfrac{1}{2}E\epsilon_1^2 = \tfrac{1}{2}\sigma_1 \cdot \epsilon_1 = \tfrac{1}{2}\sigma_1^2/E \qquad (1.20)$$

At a point in an elastic body where all the three principal stresses σ_1, σ_2 and σ_3 are non-zero the total elastic energy stored per unit volume is

$$w = w_1 + w_2 + w_3$$

and with the tensile strains given by equation 1.15 one obtains

$$w = \frac{1}{2E} [(\sigma_1^2 + \sigma_2^2 + \sigma_3^2) - 2\nu(\sigma_1 \sigma_2 + \sigma_2 \sigma_3 + \sigma_3 \sigma_1)]$$

18

Now, w consists of two parts. One, w_s, arises from deviatoric strains and hence from changes of shape, while the second, w_h, is the hydrostatic component associated with changes in size. By analogy with equation 1.20 one finds

$$w_h = \tfrac{1}{2} \sigma_h \cdot \epsilon_h \qquad (1.21)$$

with σ_h and ϵ_h given by equations 1.5 and 1.17 respectively. Thus one obtains

$$w_h = [(1 - 2\nu)/6E] (\sigma_1 + \sigma_2 + \sigma_3)^2 \qquad (1.22)$$

and, since $w_s = w - w_h$,

$$w_s = \frac{2(1+\nu)}{3E} \left[\left(\frac{\sigma_1 - \sigma_2}{2}\right)^2 + \left(\frac{\sigma_2 - \sigma_3}{2}\right)^2 + \left(\frac{\sigma_3 - \sigma_1}{2}\right)^2 \right] \qquad (1.23)$$

From equations 1.18 and 1.21

$$w_h = \frac{1}{2K} \left(\frac{\sigma_1 + \sigma_2 + \sigma_3}{3}\right)^2$$

and on comparing this with equation 1.22 the bulk modulus is obtained in terms of E and ν :

$$K = E/3(1 - 2\nu) \qquad (1.24)$$

For an incompressible material Poisson's ratio must therefore be equal to $\tfrac{1}{2}$.

Other relations between elastic constants of isotropic materials may be derived by analogous methods, applied however to w or w_s. For example, the elastic energy stored in a cube subjected to a simple shear, as in *Figure 1.7a*, is $\tfrac{1}{2}\tau_{zy}^2/G$, where the shear modulus is defined by $G = \tau_{zy}/\gamma_{zy}$. This energy can also be obtained by substituting for the principal stresses from equation 1.14 into the expression for the total stored energy derived above; the result is

$$w = \frac{1}{E} (1 + \nu)\tau_{zy}^2$$

19

Again, comparing both values, one derives the relation

$$G = E/2(1 + \nu) \tag{1.25}$$

For incompressible materials

$$G = \tfrac{1}{3} E \tag{1.26}$$

The last result may sometimes be of considerable use in practice, for example when E but not G are known, or vice versa. In the case of the more common metals, lead has a particularly low compressibility, with $\nu = 0 \cdot 44$. The elastic moduli, measured at room temperature, are $E = 1 \cdot 8 \times 10^5$ kg/cm^2 and $G = 0 \cdot 60 \times 10^5$ kg/cm^2; the ratio E/G thus agrees with equation $1 \cdot 26$ reasonably well.

The decrease of the elastic constants with increasing temperature is comparatively small and, except close to the melting point, the variation can generally be represented with sufficient accuracy for most purposes by a linear relation. If, for example, the low-temperature value of the shear modulus is known (G_0), one may write

$$G(T) = G_0\left(1 - a\frac{T}{T_m}\right)$$

where T_m is the melting temperature, expressed in °K, and a is a constant equal to about $0 \cdot 2$.

1.10 Locating the Principal Stresses in Practice

In view of the extensive reference to principal axes which we have made in the preceding sections, we shall now give a simple example on their location in a practical case. The most important consideration is generally the fact that a surface not actually subjected to shearing forces must also be free from shear stresses. A principal stress therefore acts along the normal to that surface at the point considered. Knowledge of the direction of one more principal stress then suffices to locate the axes at the point. Symmetry or other geometrical features generally provide an indication of the second required direction.

LOCATING THE PRINCIPAL STRESSES IN PRACTICE

In a thin-walled pressure vessel, which we take as our example, the internal pressure p does not result in a shear stress either in the shell or in the circular flat ends. Referring to *Figure 1.8*, we see that the principal stresses must in fact occur in axial, radial and circumferential directions. There is no tensile stress applied radially across the shell wall, and hence, in a thin-walled cylinder, $\sigma_3 = 0$. The stresses σ_1 and σ_2 are evaluated as follows.

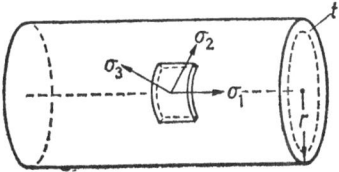

Figure 1.8. Principal stresses in the shell of a thin-walled cylinder subjected to internal pressure, p

Firstly, the force on the cylindrical shell in the axial direction is $2\pi r t \sigma_1$, where r and t are the radius and wall thickness of the cylinder. This is in equilibrium with the force $\pi r^2 p$ acting on the flat base, so that

$$\sigma_1 = pr/2t$$

Secondly, by considering the equilibrium of forces on a horizontal, axial, section through the cylinder one has for unit length:

$$2\sigma_2 t = 2rp$$

so that

$$\sigma_2 = pr/t$$

The stress system is now fully specified.

MACROSCOPIC PLASTICITY

2.1 Yield Criteria for Isotropic Materials

In uniaxial tension the onset of plastic deformation is specified by the yield stress σ_y or, if the material has been work-hardened by previous plastic deformation, by the strain-dependent flow stress σ, for example that corresponding to the point P in *Figure 1.1*. Criteria for the onset of plasticity must necessarily be more complex in the case of triaxial stress systems, although they must of course include the uniaxial deformation as a special case.

Now, the change of shape of the material resulting from the plastic deformation should depend only on the deviatoric stresses, for even extremely high hydrostatic stresses are known not to lead to permanent, plastic, deformations of ductile materials, assuming them to be homogeneous. The yield criterion would therefore be expected to include only combinations of deviatoric stresses, and therefore to have the form

$$f(\sigma_1 - \sigma_2, \sigma_2 - \sigma_3, \sigma_3 - \sigma_1) = 0 \qquad (2.1)$$

in which the deviatoric stresses given by equation 1.3 have been combined in a simple, symmetric, manner, with the concomitant elimination of σ_h. The variables of the function f can be seen to be the principal shear stresses given by equation 1.9. Reversal of the signs of all the stresses should leave the function f invariant, for it is well known, for example, that a metal rod will yield plastically in tension at the same absolute value of the flow stress as in compression. This consideration suggests that f is a quadratic function of the principal shear stresses. It should also be symmetric with respect to the three principal stresses, for none of them is in any way preferred. A relation which satisfies these requirements is

$$(\sigma_1 - \sigma_2)^2 + (\sigma_2 - \sigma_3)^2 + (\sigma_3 - \sigma_1)^2 = 2\,\sigma^2 \quad (2.2)$$

where σ is a constant the significance of which has yet to be established; the factor 2 is used for convenience later. On comparing equations 2.2 and 1.23 the criterion is seen to imply that yielding will occur when the shear energy per unit volume attains a certain definite value characterised by the constant σ. If one considers the special case of a uniaxial tensile test, putting $\sigma_2 = \sigma_3 = 0$ into equation 2.2, one finds that σ is identical with the tensile flow stress of the material. This maximum-shear-energy criterion, originally proposed by Huber, is frequently also associated with the names of Hencky and von Mises. Its applicability to ductile solids, particularly to metals, has been confirmed by extensive tests.

A second criterion, often simpler to use in practice, is known as the maximum-shear-stress criterion; it is variously associated with the names of Coulomb, Tresca, Mohr and Guest. It states that yielding will occur when the largest principal shear stress attains a certain definite level, and may be written

$$\sigma_1 - \sigma_3 = \sigma; \qquad \sigma_1 > \sigma_2 > \sigma_3 \quad (2.3)$$

As before, σ is found to be the tensile flow stress of the material. The hydrostatic stress does not appear in the criterion, but the equation is not symmetric in the principal stresses, for the intermediate stress σ_2 is taken into account only through an inequality.

The two criteria can be compared by using the parameter

$$\mu = (2\sigma_2 - \sigma_1 - \sigma_3)/(\sigma_1 - \sigma_3)$$

which varies between $+1$ and -1 as σ_2 takes values

23

between its upper and lower limits σ_1 and σ_3. Equation 2.2 can then be written

$$\sigma_1 - \sigma_3 = \sigma \ [2/(3 + \mu^2)^{\frac{1}{2}}]$$

showing that the value of $\sigma_1 - \sigma_3$ is at most by a factor $2/3^{\frac{1}{2}}$ greater than the corresponding value given by equation 2.3. The difference is therefore always less than about 16 per cent.

In the case of the thin-walled cylinder subjected to an internal pressure p, discussed in section 1.8, the maximum-shear-energy criterion shows that yielding will set in when p attains the value $\sigma t/3^{\frac{1}{2}}r$, while equation 2.3 yields $\sigma t/2r$; in this case $\mu = 0$. If it is desired to avoid plastic deformation of the cylinder then the maximum-shear-stress criterion will give a lower and hence 'safer' estimate of the permitted maximum pressure. In view of its simplicity it is frequently used in practice although equation 2.2 is generally found to agree somewhat better with experiment.

2.2 Yielding of Single Crystals

Plastic yielding in single crystals occurs by slip, sometimes also termed glide, in which thin lamellae of the crystal glide over neighbouring ones like playing cards in a packet. In metals the crystallographic slip planes which are preferred, i.e. along which slip takes place under the smallest shear stresses, are generally those most densely studded with atoms, and the directions of slip are along the shortest lattice spacings between crystallographically equivalent sites. In aluminium, copper, gold, lead, silver and many common metals having the face-centred cubic structure, slip takes place preferentially on the eight sets of ' octahedral ' planes, one of which is shown in *Figure 2.1a*. There are six possible slip directions in each plane, i.e. two along each face diagonal of the triangle shown. Altogether there are therefore 24 possible slip directions: two along each of the 12 face diagonals. In simple ionic crystals slip directions lie generally along the joins of nearest neighbours of equally charged ions; *Figure 2.1b* shows a slip plane in a simple cubic crystal of the rocksalt type.

Glide occurs when the shear stress acting in the slip direction of a preferred system of glide planes attains a definite value, known as the critical resolved shear stress τ_{cr}. If, referring to *Figure 1.2*, a cylindrical crystal is subjected to uniaxial tension, and the angle between the normal of a glide plane and the tension axis is θ, while the glide

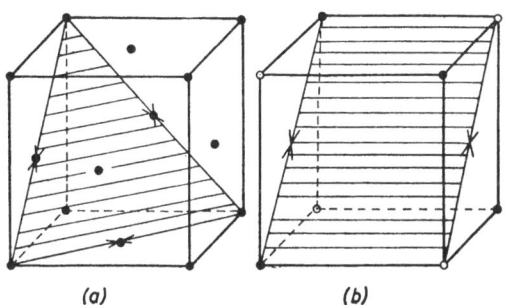

(a)   (b)

Figure 2.1. Preferred slip planes in: (a) face-centred cubic metals; (b) simple ionic crystals of the rocksalt type. Arrows indicate slip directions

direction makes an angle ϕ with the axis, then the component of the force along this direction is $F \cos \phi$. As the area of the glide plane is $A/\cos \theta$ the resolved shear stress in the glide direction is

$$\tau_{cr} = \sigma \cos \theta \cos \phi \qquad (2.4)$$

where $\sigma = F/A$. If several preferred glide systems exist, as for example in crystals of the type referred to in *Figure 2.1*, then on gradually increasing the tensile stress from zero a value of σ will be attained which will satisfy equation 2.4 for at least one of the slip planes. In fact the slip system with the highest value of the product $\cos \theta . \cos \phi$ will become operative first. There is extensive evidence in support of the critical shear stress law; its validity is readily checked experimentally by examining the constancy of this product with crystals variously oriented with respect to the tensile axis. Attempts have been made to show that equation 2.4 may be generalised so as to lead to yield criteria which are known to hold for polycrystalline aggregates, i.e. equations 2.2 or 2.3.

2.3 Equations of Plasticity

As we have already noted, a wire extended beyond the yield point acquires a permanent, plastic, deformation, and its length does not revert to its initial value if the stress is removed. Although the wire is now strained the applied stress is zero. Clearly, in view of this irreversibility of deformation, a linear relation between stress and strain, akin to Hooke's Law, cannot hold in the plastic range. However, we know that the previous stress would have to be applied, and in fact slightly exceeded, if we wished to extend the wire further by a small amount. One could therefore expect a relation between stresses and increments of plastic strain. In place of Hooke's Law, referred to principal axes, one may then write three differential equations.

$$\left. \begin{array}{l} d\,\epsilon_{p1} = dC.[\sigma_1 - \tfrac{1}{2}\,(\sigma_2 + \sigma_3)] \\ d\,\epsilon_{p2} = dC.[\sigma_2 - \tfrac{1}{2}\,(\sigma_3 + \sigma_1)] \\ d\,\epsilon_{p3} = dC.[\sigma_3 - \tfrac{1}{2}\,(\sigma_1 + \sigma_2)] \end{array} \right\} \qquad (2.5)$$

where C represents a yet undefined, strain-dependent, property of the material, and the subscript p indicates a plastic component. As irreversible deformations occur in most materials without significant volume changes, Poisson's ratio is taken equal to $\tfrac{1}{2}$, as required by equation 1.24.

Now we notice that

$$\sigma_1 - \tfrac{1}{2}\,(\sigma_2 + \sigma_3) = \tfrac{3}{2}\,(\sigma_1 - \sigma_h)$$

26

where σ_h is the hydrostatic stress given by equation 1.5. Analogous expressions hold for the other bracketed terms in equation 2.5. On writing σ_{d_1} for the deviatoric stress component $\sigma_1 - \sigma_h$, with a corresponding notation for $\sigma_2 - \sigma_h$ and $\sigma_3 - \sigma_h$, equation 2.5 can be recast into the form

$$\frac{d\epsilon_{p_1}}{\sigma_{d_1}} = \frac{d\epsilon_{p_2}}{\sigma_{d_2}} = \frac{d\epsilon_{p_3}}{\sigma_{d_3}} = \tfrac{3}{2}\,dC \qquad (2.6)$$

first derived by Prandtl and Reuss. The significance of dC can be established from the following considerations. The 'plastic' work per unit volume of solid due to simultaneous increments of the plastic strain is

$$dw_p = \sigma_{d_1}\,d\epsilon_{p_1} + \sigma_{d_2}\,d\epsilon_{p_2} + \sigma_{d_3}\,d\epsilon_{p_3}$$

which, in view of equation 2.6, may be written

$$dw_p = \tfrac{3}{2}\,dC\,(\sigma_{d1}^2 + \sigma_{d2}^2 + \sigma_{d3}^2) \qquad (2.7)$$

It may readily be verified that equation 2.2 is equivalent to

$$\sigma_{d1}^2 + \sigma_{d2}^2 + \sigma_{d3}^2 = \tfrac{2}{3}\,\sigma^2$$

so that

$$dw_p = \sigma^2 \,.\, dC \qquad (2.8)$$

in which the individual stress components do not appear explicitly. We may therefore consider the special case of a rod, or wire, extended plastically by uniaxial tension, the yield stress being σ. The increment of 'plastic' work is $\sigma.d\epsilon_{p_1}$, so that equation 2.8 yields

$$d\epsilon_{p_1} = \sigma \,.\, dC$$

and on writing for the coefficient of work-hardening, given by the slope of the stress-strain curve at the flow stress σ,

$$H = d\sigma/d\epsilon_{p_1} \qquad (2.9)$$

one obtains

$$dC = d\sigma/H\sigma \qquad (2.10)$$

27

Thus if the yield or flow stress σ and the corresponding coefficient of work-hardening are known, it is possible to determine the strain increments resulting from a small increase of σ, as is apparent from equations 2.6 and 2.10. Integration of equation 2.6 is not however in general possible.

As a specific, simple, illustration we may consider the case where $H = \chi/\sigma$, χ being a constant, and $\sigma_1 \equiv \sigma$ is the only non-zero stress. Equations 2.6 and 2.10 then yield

$$d\epsilon_{p1} = (\sigma_1/\chi)d\sigma_1$$

and

$$\frac{d\epsilon_{p2}}{-\tfrac{1}{3}\sigma_1} = \frac{d\epsilon_{p3}}{-\tfrac{1}{3}\sigma_1} = \frac{d\epsilon_{p1}}{-\tfrac{2}{3}\sigma_1}$$

The solution

$$\sigma_1^2 = \chi\epsilon_{p1} \; ; \; \epsilon_{p2} = \epsilon_{p3} = -\tfrac{1}{2}\epsilon_{p1}$$

shows that the wire, or rod, hardens 'parabolically', and also that the sum of the three plastic strain components is zero. The last result could of course have been expected because of the assumed incompressibility of the material (equation 1.19).

2.4 Approximate Methods and Applications

The mode of plastic deformation in wire drawing, deep drawing of metal sheet, extrusion, rolling and other shaping processes is generally too complex to yield to rigorous analysis, but results useful in practice may still be obtained if reasonable simplifying assumptions are made. The large deformations which are as a rule involved also call for a redefinition of strain; clearly it would not be convenient or physically meaningful to express, let us say, the plastic strain of a long thin wire in terms of the initial dimensions of the relatively small metal billet from which it was drawn.

Again consider a rod of initial cross-section A_1 and length l_1 ; the material is assumed incompressible. As its volume is therefore constant one has

$$A_1 l_1 = A_2 l_2 = Al = V = \text{constant} \qquad (2.11)$$

28

where A is the cross-section when the rod has been extended to length l, and A_1 and A_2 refer to the initial and final states respectively. The flow stress σ at any stage of the deformation will depend upon the plastic strain, as is clear from *Figure 1.1*. The 'plastic' work expended in the deformation is

$$W_p = \int_{l_1}^{l_2} \sigma(l) \, . \, A \, . \, dl = V \int_{l_1}^{l_2} \sigma(l) \frac{dl}{l} \qquad (2.12)$$

the second integral being obtained by the use of equation 2.11. Hence the work per unit volume, W_p/V, is

$$w_p = \int_{l_1}^{l_2} \sigma \, (l) \, . \, d \, (\ln l) \qquad (2.13)$$

On comparing equations 2.12 and 2.13 with corresponding expressions for the work done in elastic deformations, given in section 1.9, one finds that the tensile strain is now replaced by the natural logarithm of the instantaneous length of the rod. This so-called 'natural strain', sustained by the rod on being extended from length l_1 to l_2, is $\ln l_2 - \ln l_1$, or $\ln (l_2/l_1)$. This may also be written $\ln[1 + (\Delta l/l_1)]$ where $\Delta l = l_2 - l_1$. If $\Delta l \ll l_1$ the logarithm may be expanded as a power series, and terms of second and higher degrees in $\Delta l/l_1$ may be neglected. The natural strain is then found to be equal to the conventional strain $\Delta l/l_1$.

Equation 2.13 also implies that the area under the stress versus natural strain curve, bounded by two values of the strain, is numerically equal to the work per unit volume expended in straining the rod from length l_1 to l_2. It is sometimes convenient to replace the actual stress–strain curve by one representing an 'ideally' plastic material, as shown in *Figure 2.2*.

If the level of the yield stress of the ideally plastic body is judiciously chosen, problems in plasticity may become mathematically simplified without undue sacrifice of precision; this applies particularly at strains where the slope of the work-hardening curve is relatively small.

Taking σ_0 to be the yield stress of the ideally plastic solid used to replace the actual one (*Figure 2.2*), equation 2.13 reduces to the simple relation

$$w_p = \sigma_0 \ln (l_2/l_1) = \sigma_0 \ln (A_1/A_2) \qquad (2.14)$$

The form involving the areas follows from the constancy of volume as expressed by equation 2.11.

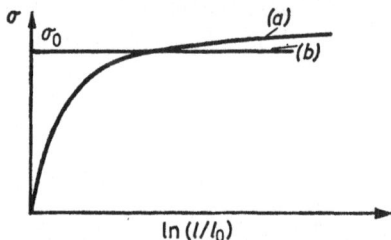

Figure 2.2. (a) The actual natural stress-strain curve; (b) its approximation by an ideally plastic body

The rod could also have been extended by extrusion, drawing through a die, or by rolling. In such operations the useful work necessary to impart the required change of shape is always accompanied by some wasted 'plastic' work, so that the efficiency of the process, ψ, defined as the ratio of useful to total work expended, is less than 1. The origin of some of the wastage of work can be seen by considering extrusion. As is apparent from *Figure 2.3*, a short horizontal element of the material in the billet container is first bent into a convex shape and subsequently straightened as it passes through the die or 'virtual die' formed by the static 'dead' material at the container wall around the exit orifice. The bending requires expenditure of 'plastic' work without contributing to the required change of shape.

30

Thus apart from frictional effects at the billet–container interface, which can often be made relatively unimportant by appropriate lubrication, allowance has to be made for this. Equation 2.14 has then to be modified to

$$w_p = (\sigma_0/\psi) \ln(A_1/A_2) \qquad (2.15)$$

Values of ψ generally lie between $0 \cdot 5$ and $0 \cdot 8$ for most simple forming processes.

The force F_1 which has to be applied to the ram (*Figure 2.3*) to effect extrusion can be evaluated as follows. Neglecting billet–container friction, and assuming an ideally plastic material we see that the work $F_1 x$ done in displacing the

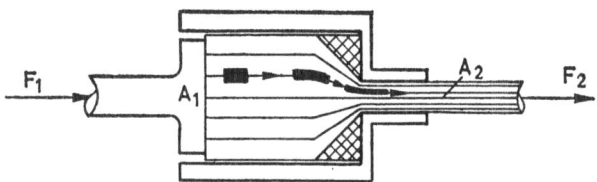

Figure 2.3. Extrusion v fo rod of cross-section A_2 from a billet of cross-section A_1. Reversed bending of an element indicates expenditure of 'useless' work

ram a distance x towards the die leads to the extrusion of a volume $A_1 x$ of material, and consequently to an expenditure of work $A_1 x w_p$. On equating this to $F_1 x$, and using equation 2.15, one obtains

$$F_1 = A_1 (\sigma_0/\psi) \ln(A_1/A_2) \qquad (2.16)$$

If the material is drawn through a die by a force F_2 instead then similar reasoning shows that

$$F_2 = (A_2/A_1)F_1$$

where F_1 is given by equation 2.16. The stress on the material due to the applied drawing force F_2 is equal to about $(\sigma_0/\psi) \ln (A_1/A_2)$, and this must be less than σ_0 if the wire, or rod, is not to break. Hence an upper limit is set to the possible reduction in area per pass.

The preceding principles are readily adapted to other shaping processes, such as draw rolling—in which metal sheet is pulled by a coiler drum through free-wheeling rolls —or to deep drawing. In the former case, *Figure 2.4*, the

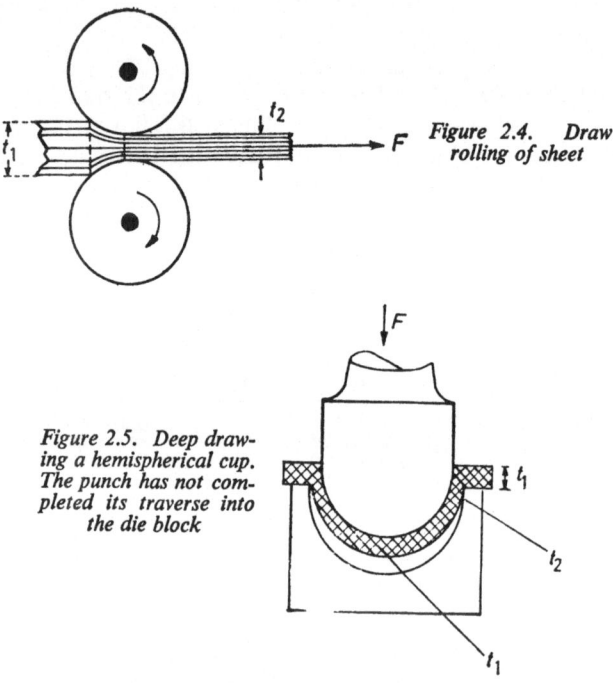

Figure 2.4. Draw rolling of sheet

Figure 2.5. Deep drawing a hemispherical cup. The punch has not completed its traverse into the die block

analogy with extrusion is obvious, while in deep drawing, *Figure 2.5*, it may be discovered as follows.

We consider a hemispherical metal cup drawn from a circular blank of equal diameter. In the course of being drawn it assumes intermediate shapes, as shown shaded in the figure. The thickness at the rim is equal to the final width, t_2 but at the centre of the base the thickness is still the same as that of the blank. The effective die may therefore be considered to extend at any time during the drawing process over the parts of the cup where the thickness lies between the initial and final values, t_1 and t_2. The area of the part of the blank converted into the cup is equal to $\frac{1}{2}$ that of the cup, so that

$$w_p = (\sigma_0/\psi) \ln 2$$

As the punch travels into the die-block a total distance approximately equal to the radius r of the cup one has, assuming uniformity of deformation,

$$F \cdot r = V \cdot w_p$$

where V is the volume of a circular area of the blank of radius r, i.e. $\pi r^2 t_1$. Thus

$$F = (\pi r t_1 \sigma_0/\psi) \ln 2$$

In hot shaping operations, when the temperature of the material is in excess of about $0 \cdot 4 T_m$ (°K), where T_m is the melting temperature, the flow stress generally increases significantly with increasing strain rate, an effect which will be discussed later in connection with high-temperature creep. A knowledge of the strain rate may therefore be important in the determination of ram pressures, press capacities, and so on. We shall indicate here how the strain rate may be estimated in the case of extrusion; similar methods appropriate to other processes suggest themselves readily.

In the corner of the container, *Figure 2.3*, an accumulation of static ' dead ' material forms a virtual die in the shape of a truncated, approximately rectangular cone; the planes of maximum shear stress forming the cone surface in general make an angle of about 45° with the billet axis. Deformation of the material occurs mainly within this die. Now the passage of a small volume of material, initially at the die mouth, through the die will take place over an interval of time given by the ratio of die volume to volume of material extruded per unit of time. On taking the die height to be approximately $\frac{1}{2}D$, where D is the billet diameter, the time of passage is found to be

$$t = (\pi D^3/24)/(\pi D^2 v/4) = D/6v$$

where v is the ram velocity. The strain rate is therefore

$$\frac{d\epsilon}{dt} = \frac{12v}{D}\ln\frac{D}{d}$$

since $\ln(A_1/A_2) = 2\ln(D/d)$, d being the diameter of the extruded rod.

2.5   The Stress–Strain Curve

The mechanical behaviour of materials is generally investigated by tests each of which is characterised by the use of a specific stress system, such as tension, compression, torsion, bending, or a combination of these, as well as by the time rate of application of the stresses and the total time under load. Thus if the time rate is zero, the material being subjected to constant stresses for long periods, we have conditions used in studies of creep; bending with impact loading occurs in studies of brittleness by the notched-bar method, while prolonged application of periodically varying stresses is employed in the investigation of fatigue. At present we shall confine our attention to the tensile test, as an example of a widely used ' static ' method of studying the plastic behaviour of materials. The principles involved in the measurement of hardness, impact strength and toughness, in all of which loading is relatively rapid, i.e. ' dynamic', we shall briefly examine later.

In a tensile test a specimen in the form of a rod, wire or strip, is subjected to gradually increasing loads and the relation between the tensile force and the extension is recorded. The machine may extend the test piece at a constant velocity, and the strain rate is then roughly constant in the course of extension; alternatively a nearly

34

constant rate of stressing may be obtained by pouring
liquid or granular material into a loading pan or bucket
suspended from the specimen. In the first case the machine
is said to be 'hard', the second arrangement is termed
'soft.' Few machines are either purely hard or soft, for
it is in practice rather difficult to achieve strict control
over the rates of strain or stress.

The well-known 'yield-drop' of the load which is
observed with certain alloys, such as mild steel, and brasses

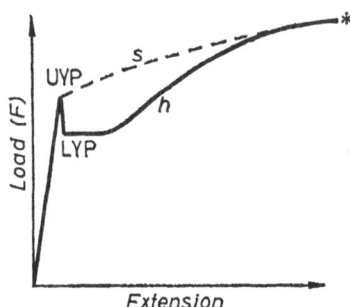

Figure 2.6. Load–extension curves for a
material with a pronounced yield effect,
obtained on a hard (h) and soft (s) tensile testing machine

containing 20–30 per cent by weight of zinc, indicated in
Figure 2.6, could not occur with a soft machine, for a drop
in the applied load is precluded by the constancy and
positive value of the loading rate. The characteristics of
the machine may therefore influence the load–elongation
curve appreciably under certain circumstances.

A typical load–elongation curve, such as may be obtained
with a great variety of materials, is shown in *Figure 2.7*.
Provided the material is ductile, so that fracture, indicated
by a star, does not occur at too small a strain, the curve is
found to become convex, with a definite maximum at a

load F_{max}. This maximum load, divided by the cross-section of the undeformed test piece, A_0, is referred to as the ultimate tensile strength, σ_{ult}, often also denoted by UTS.

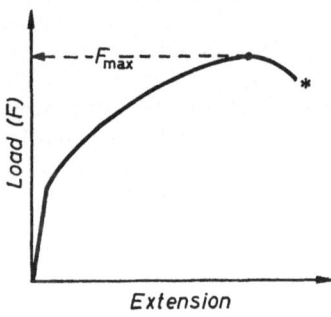

Figure 2.7. Typical load–extension curve of a ductile material, with a maximum attained before fracture

Its occurrence is a consequence of the method of testing; as we shall show it does not correspond to any sudden changes in the physical properties of the material, although it may provide a practical measure of the load-bearing capacity of a metal rod, for example.

By definition we have

$$\sigma_{ult} = F_{max}/A_0 \qquad (2.17)$$

while the corresponding true stress, referred to the actual cross-section A_{ult} is

$$\sigma_{max} = F_{max}/A_{ult} \qquad (2.18)$$

If the true stress corresponding to an instantaneous rod cross-section A is σ, then

$$F = A\sigma$$

The force will attain its maximum F_{max} when

$$d(A\sigma) = 0 \qquad (2.19)$$

i.e. when

$$- \, dA/A = d\sigma/\sigma$$

36

As we are assuming the material to be incompressible it follows from equation 2.11 that $-\,dA/A = dl/l$, where l is the instantaneous length of the specimen. We may therefore write

$$d\sigma/\sigma = (dl/l_0)\,(l_0/l)$$

or

$$d\sigma/\sigma = \frac{d\,(l - l_0)}{l_0} \cdot \frac{l_0}{l_0 + \Delta l}$$

which yields

$$\frac{d\sigma}{d\epsilon} = \frac{\sigma}{1 + \epsilon} \tag{2.20}$$

As equation 2.19 was used in deriving this result, $d\sigma/d\epsilon$ is the slope of the stress–strain curve where, by equation 2.18, the stress is equal to σ_{max}. Hence σ_{max} may be found on the

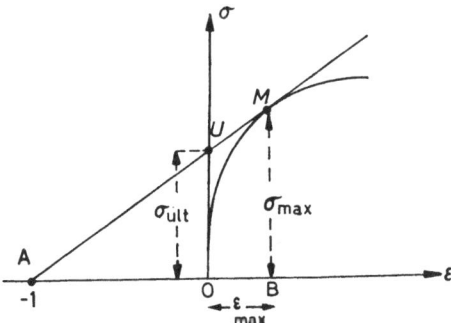

Figure 2.8. Determination of the ultimate tensile strength from the true stress-strain curve

true stress–strain curve as the point M of contact of the tangent to it drawn from the point $\epsilon = -\,1$ on the strain axis. This is apparent from *Figure 2.8*. The magnitude of the ultimate stress is then obtained from the relation

$$\sigma_{ult}/\sigma_{max} = A_{ult}/A_0 = l_0/l = 1/1 + \epsilon_{max}$$

37

which follows from equations 2.17 and 2.18. The geometrical significance of this equation can be seen by considering the similar triangles ABM and AOU in *Figure 2.8*; clearly σ_{ult} is represented by the intercept OU which the tangent makes on the stress axis. The existence of a maximum in the force–elongation curve is therefore seen to result from the failure to allow for the continuous decrease of the test piece cross-section in the course of deformation in that representation.

In practice the UTS provides an indication of the stress at which local constrictions, also termed 'necks,' begin to form in the test piece. Deformation then tends to occur preferentially at these points, as the tensile stress is there above the average for the specimen. The material may tend to draw down into a double cone; in ductile metals a characteristic cup-and-cone fracture then often results. The strain at fracture or the reduction in cross-section at the neck at which failure occurred are used as criteria of ductility in practice.

2.6 Hardness and Toughness

The resistance of plastic materials to deformation is frequently assessed by their hardness, which is also an important, though by no means sole, parameter determining their resistance to abrasion and wear. Most measurements of hardness are based on the determination of the area of an indentation made by a pyramidal diamond (Vickers, Knoop) or spherical metal indentor (Brinell, Rockwell) under a definite load, which is allowed to act for a specified time ranging from a fraction of a second to a few seconds.

We shall discuss the Brinell test as a typical example of the use of a spherical indentor, in this case a hardened steel sphere, and the Vickers test in which a diamond pyramid with a square base is used.

The Brinell hardness number (BHN) is defined as the stress W/A kg/mm^2, where W is the load applied to the indentor, and A the surface area of the spherical cap forming the indentation. If D and d are the indentor and indentation diameters respectively, one has

$$\text{BHN} = W/(\tfrac{1}{2}\pi D^2)\left\{1 - [1 - (d/D)^2]^{\frac{1}{2}}\right\}$$

or

$$\text{BHN} = [W/(\tfrac{1}{4}\pi d^2)]\,\text{f}(d/D) \qquad (2.21)$$

where

$$\text{f}(d/D) = \tfrac{1}{2}\,(d/D)^2/\left\{1 - [1 - (d/D)^2]^{\frac{1}{2}}\right\}$$

Now, if the indentation is small, so that $d \ll D$, then expansion of the term in square brackets and neglect of second and higher order terms in d/D yields $\text{f}(d/D) = 1$, while with large indentations, when $d/D \approx 1$, $\text{f}(d/D) \approx \tfrac{1}{2}$. It follows that consistent results irrespective of indentor diameter can be obtained only if the ratio d/D is maintained constant, and preferably small, the latter requirement being desirable to facilitate ready conversion of the BHN to other hardness numbers, as will become apparent below. In practice the indentor is kept under load for about 20 seconds, and the diameter of the identation is then measured by means of a low-power microscope. The result is used to obtain the BHN directly from tables.

In the Vickers 'pyramid hardness test' a hydraulic mechanism applies a load W for a fraction of a second to a diamond indentor initially arranged so as to touch the surface of the material with its tip. The pyramid has a square base, and opposite faces make an angle of 136° with one another. The use of a large angle helps to minimise friction effects.

If one assumes that a uniform normal pressure p acts on the indentor, then by considering a face, such as AOB in *Figure 2.9*, the vertical component of force opposing the load on this face is seen to consist of contributions $2xdl\,.\,p\,.\,\sin\alpha$ and $2xdl.\mu p.\cos\alpha$, where μ is the coefficient of friction at the diamond–material interface, and dl the width of the elementary strip a distance x along the face from the apex. In equilibrium

$$W = 4p \int_{x=0}^{\frac{1}{2}a} (\sin \alpha + \mu \cos \alpha) . 2x . \mathrm{d}l$$

or

$$W = p (1 + \mu \cot \alpha) \int_{x=0}^{\frac{1}{2}a} 8x . \mathrm{d}x$$

The integral yields the projected area of indentation, a^2, and the VHN, given by W/a^2, is then

$$\text{VHN} = p (1 + \mu \cot \alpha) \qquad (2.22)$$

With the large values of the cone semi-angle used the friction term $\mu \cot \alpha$ is generally negligible compared with 1, and the VHN is then equal to p. In practice the indentation diagonal is measured, and the VHN is then read directly from tables. It can now be seen that the BHN and VHN should be equal for a given material, provided $f(d/D)$ in equation 2.21 is close to 1, i.e. $d/D \ll 1$. Similarity requirements such as relate to $f(d/D)$ do not arise with pyramidal indentors.

Most materials become harder and less ductile as the temperature is lowered; well known examples include sealing wax, rubber and metals. A piece of rubber submerged in liquid nitrogen (77°K) and then withdrawn can be shattered with a hammer; an abrupt loss of ductility also occurs in most common types of steel, generally in the range 0 to -40°C. Resistance to impact loading of embrittled materials is low; fracture is readily initiated if notches, scratches and other stress-raising defects occur in the material. Sometimes notch sensitivity of this type is useful, for example on cutting sheet glass with diamond tipped tools.

An embrittled material will fracture without significant expenditure of work on plastic deformation; in fact the work expended in fracturing standard test pieces provided with rectangular notches of standard depth and root radii is used in the Charpy and Izod tests for comparing toughness or, as it is sometimes called, ' notch brittleness ' of materials.

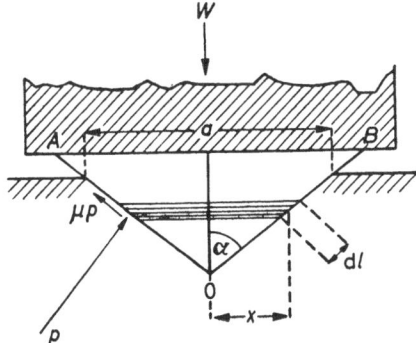

Figure 2.9. Indentation of an ideally plastic material by a diamond pyramid indentor. AOB is one of the four indentor faces

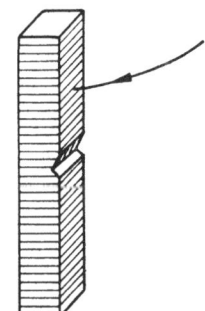

Figure 2.10. A ' Charpy ' V-notch test piece for notch impact strength determinations. Depth of notch is 2 mm, *notch root radius is* 0·25 mm

Figure 2.10 shows a Charpy V-notch specimen used for impact testing. The sides of the square base measure 1 cm, and the notch, situated at the centre of the 5·5 cm long specimen, is 2 mm deep. The notch root radius is 0·25 mm.

The bottom part of the specimen is securely clamped and the top is struck from the notch side by a pendulum hammer, which loses energy in fracturing the specimen. The difference between the down and up-swing angles of the pendulum, read from a scale, yields the energy absorbed in fracturing the specimen. It is expressed in kg m and provides a measure of the notch impact strength, useful in particular in investigating the effect of temperature changes, heat treatments and alloying, on the toughness of the material.

PLASTICITY AND STRENGTH OF CRYSTALS

3.1 The Theoretical Strength of Crystals

The shear stress at which plastic deformation should begin to take place in an ideally perfect crystal was first calculated by Ya. Frenkel in 1926. The method is straightforward. Referring to *Figure 3.1*, which represents two adjacent planes of atoms of a crystal of simple structure, we see that

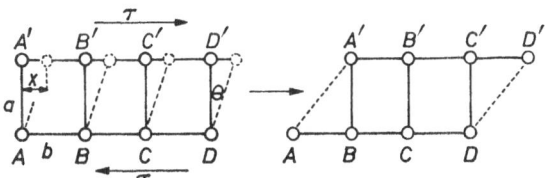

Figure 3.1. Slip in an ideally perfect crystal

if the upper one were displaced relative to the lower one in the slip direction *b* so that the atoms *A'*, *B'*, *C'*, etc. came to lie over *B*, *C*, *D*, etc. respectively, slip would have taken place. The crystal would remain crystallographically perfect as the atoms have moved to equivalent lattice sites. The shear stress necessary to produce this permanent plastic deformation is the critical shear stress τ_c of the ideally perfect crystal.

To obtain its value we first consider the form of the stress-displacement curve (*Figure 3.2*), measuring displacements *x* indicated in *Figure 3.1*. Now $\tau = 0$ when $x = 0$ and $x = b$; the crystal is in stable equilibrium in both cases.

When $x = \frac{1}{2}b$ the atoms in the upper layer can be seen to be in a position of unstable equilibrium; the force in the slip direction is zero, for each atom is equally attracted to the right and to the left. Consequently $\tau = 0$ also at $x = \frac{1}{2}b$. Further, we know from the elastic behaviour of crystals that the shear stress increases linearly with the strain within

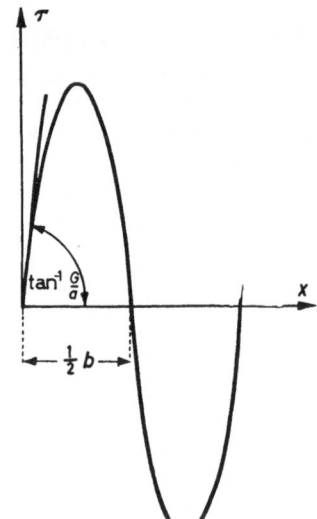

Figure 3.2. The harmonic stress versus displacement law for slip in an ideally perfect crystal

the elastic range. The slope of the shear stress versus displacement curve at $x = 0$ and $x = b$ must therefore comply with Hooke's Law

$$\tau/\gamma = G \qquad (3.1)$$

where

$$\gamma = x/a$$

Thus

$$d\tau/dx = G/a, \qquad x \ll a \qquad (3.2)$$

Now $\tau(x)$ must be a periodic function with wavelength b, and we therefore take the relation

$$\tau = K \cdot \sin(2\pi x/b) \qquad (3.3)$$

shown in *Figure 3.2*. Other harmonic functions satisfying the boundary conditions at $x = 0$, $\frac{1}{2}b$ and b could have been chosen, without however affecting the result significantly. As equation 3.3 must reduce to equation 3.1 for small strains, one finds that then

$$\tau = Gx/a = K(2\pi x/b)$$

so that

$$K = Ga/2\pi b \qquad (3.4)$$

The highest shear stress occurs when

$$2\pi x/b = \frac{1}{2}\pi$$

i.e. at $x = \frac{1}{4}b$, and is numerically equal to K. In practice a and b will not differ from one another appreciably, and one therefore obtains the result that the critical shear stress of an ideally perfect crystal is

$$\tau_c \approx G/10 \qquad (3.5)$$

The maximum 'elastic' shear strain also occurs when x is $\frac{1}{4}b$, and is therefore equal to $\frac{1}{4}b/a$ which, with $a \approx b$, yields

$$\gamma_c \approx 25 \text{ per cent} \qquad (3.6)$$

Crystals of the high degree of perfection necessary to check this theory can be obtained readily only in the form of microscopically thin 'whiskers'. They have nevertheless permitted sufficient experimentation to establish the correctness of the foregoing results. However, ordinarily the critical shear stress of, say, pure copper crystals is about 20 kg/cm², while $G/10$ (equation 3.5) yields about 40,000 kg/cm²; the strength of the ideal crystal is therefore 2,000 times higher than that of a real one. Similar ratios are obtained with other metallic and non-metallic crystals. The disagreement persists with polycrystalline materials although the ratio of ideal strength to true shear strength is then somewhat less than the corresponding fraction for single crystals.

The origin of this extreme weakness of real crystals, compared with ideal ones, was the subject of extensive speculation over a period of several years, but remained obscure until 1934–35 when G. I. Taylor, E. Orowán, M. Polányi, as well as Ya. Frenkel jointly with T. Kontorova, in independent papers laid the foundations for a rational theory of crystal plasticity. The common feature of these developments was the hypothesis that transport of matter, giving rise to plasticity, occurred through the agency of specific defects of the crystal lattice termed 'dislocations'. Direct evidence of the existence and behaviour of dislocations was not obtained until two decades later, when etch-pit studies and electron transmission microscopy methods confirmed the principal characteristics of dislocations which had until then been inferred mainly from theoretical considerations. In the following section we shall outline the features of the Theory of Dislocations essential to an understanding of crystal plasticity.

3.2 Edge and Screw Dislocations

An interesting analogy illustrating the role of dislocations in slip, suggested by N. F. Mott, is based on the introduc-

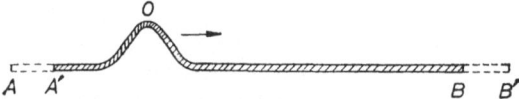

Figure 3.3. Illustration of the function of an edge dislocation. Movement of the ruck, O, transports matter from left to right

tion of a ruck, *O*, into a long runner carpet which it is desired to shift from its initial position *AB* to a new one *A' B' (Figure 3.3).*

46

The ruck, which carries the excess material originally at *AA'*, is readily impelled by a small force to move in the direction indicated, leading to the displacement of the entire carpet which, by pulling at *B*, would have been hard to move bodily. A similar ruck may be introduced into an elastic cylinder by slitting it axially up to some line *OO'* (*Figure 3.4*) and then re-sealing it with the adjacent faces out of alignment by an amount *AA'*. The line *OO'* at which the ruck is located represents an edge dislocation in the elastic cylinder.

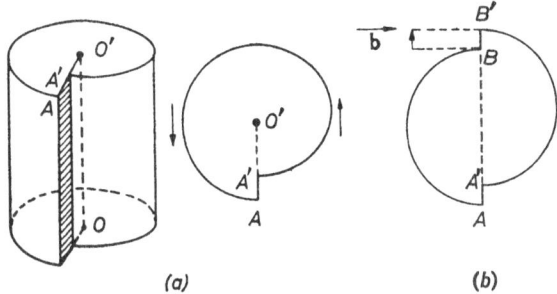

Figure 3.4. An edge dislocation of Burgers vector **b**: (a) inside an elastic cylinder ; (b) slip due to its passage through the cylinder

If the cut were extended so that the edge *OO'* moved further into the material the dislocation could move right through the cylinder under the action of a shear stress τ acting in the slip plane *AO'O*, and both halves would become displaced with respect to one another by an amount *AA*, as is apparent from *Figure 3.4b*. The cylinder is therefore sheared as a result of the passage of the dislocation through it. *AA'* (or *BB'*) is known as the Burgers vector.

In a homogeneous elastic material, such as rubber, the length *AA'* of the slip vector may be chosen arbitrarily, but in crystalline substances the directions and lengths must be those of a definite lattice spacing, as was already pointed out in section 2.2. *Figure 3.5* shows the passage of an edge dislocation lying at right-angles to the plane of the paper through a crystal.

Remember that the crystal is three-dimensional; *Figure 3.5* shows that the dislocation may be considered to result from the insertion of an extra plane of atoms up to half way down the crystal, its edge terminating in the slip plane along which shearing occurs. In the figure the extra half plane and the slip plane immediately adjacent to the edge,

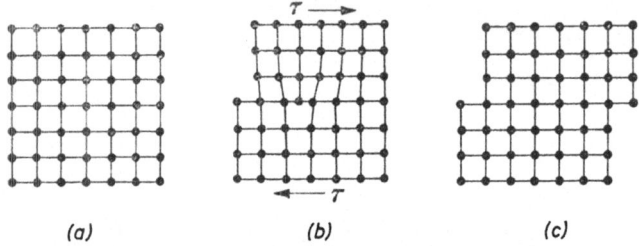

(a)   *(b)*   *(c)*

Figure 3.5. Passage of an edge dislocation through a crystal: (a) perfect crystal; (b) dislocated crystal; (c) sheared, dislocation-free crystal

i.e. at the centre of the dislocation, are seen to form an inverted letter *T*, and the symbol ⊥ is generally used to represent a positive edge dislocation. The sign, according to convention, is negative if the half-plane is inserted into the crystal from below; the appropriate symbol is then T.

Atoms adjacent to the edge, such as those along lattice lines below the slip plane are held in position relatively weakly and, in view of the looseness of the lattice in their neighbourhood, they will readily move into alignment below the centre of the dislocation. Now if, under an applied shear stress, the nearest line is drawn below the centre of the edge, then lower lines will follow, and the half-plane is converted to a normal, complete, plane of atoms. Although the dislocation thereby disappears from its initial position, transfer of the lower lines of atoms has now left a half-plane of atoms adjacent to its previous location; the process is therefore the mechanism by means of which the dislocation propagates and transports material. By contrast with the Frenkel process, slip now propagates through the crystal in stages; it does not occur by simultaneous displacement of all atoms on the operative slip plane.

In the case of a straight edge dislocation normal to the crystal face (*Figure 3.5b*) the stress necessary to induce its migration, and hence slip, known as the Peierls–Nabarro stress, can be shown to be given by

$$\tau_{PN} = [2G/(1 - \nu)] \exp(- 4\pi\zeta/b) \qquad (3.7)$$

where ζ, the half-width of the dislocation, is equal to the distance from the centre of the dislocation, measured along the slip plane, up to the point where the displacement of the atoms falls to one half of its maximum value. In the simple cubic lattice shown in *Figure 3.5*, $\zeta = a/2(1 - \nu)$ where a is the lattice spacing. As this is equal to **b**, the Burgers vector, then with $\nu = \frac{1}{3}$ the exponent in equation 3.7 is equal to about $- 9$; for face-centred cubic metals one obtains approximately the same value. For copper, in particular, τ_{PN} is found to be close to 150 kg/cm^2, which is still several times higher than the experimentally determined critical shear stress of well annealed copper crystals. Refinements of the Peierls–Nabarro treatment have been proposed, and these lead to a significant reduction of the calculated critical shear stress, although even then the discrepancy is not satisfactorily resolved. However, direct observation of dislocations by transmission electron microscopy has shown that, contrary to the assumptions made in the Peierls–Nabarro theory, dislocations are not confined to specific lattice directions, but generally lie across rather than in the potential valleys in the crystal. A stress appreciably less than τ_{PN} as given by equation 3.7 should suffice to move such dislocations.

It should be noted that the atoms at the edge of the extra plane shown in *Figure 3.5b* have fewer nearest neighbours than other atoms. In covalent crystals, such as germanium or silicon, unsaturated ' dangling ' bonds must then exist at the edge, and the dislocations will therefore

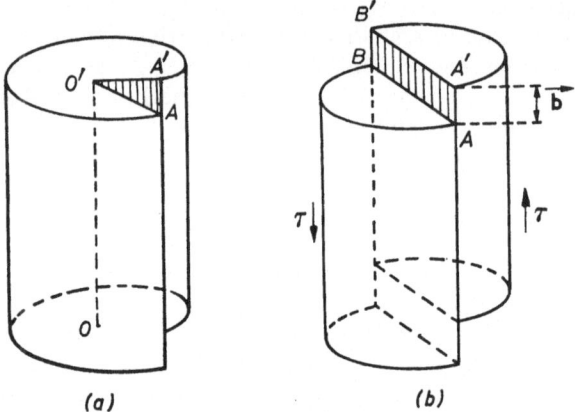

(a)   (b)

*Figure 3.6. A screw dislocation of Burgers vector **b**: (a) inside an elastic cylinder; (b) slip due to its passage through the cylinder*

be electrically charged. Impurity charge carriers of opposite sign will as a result be attracted to the dislocation and, if they can diffuse through the lattice, they will migrate towards the dislocation and attach themselves to the free bond.

Another type of dislocation of considerable importance in the growth of crystals from the vapour phase and in plasticity is known as the ' screw ' dislocation. In order to examine the geometry of such a dislocation we again cut an elastic cylinder as before, but we now displace the adjacent faces of the cut vertically with respect to one another, as shown in *Figure 3.6*. If the dislocation *OO′* is forced to pass through the cylinder from front to back, by the applica-

50

tion of a shear stress as indicated, one half will be displaced by an amount equal to the Burgers vector along AA' with respect to the adjacent one. If a complete circle is drawn around the dislocation line OO', going clockwise along the rim and starting at A, it will terminate at A', a distance b above the starting point. The path is in fact helical rather than circular, as on a screw thread; the name of the dislocation refers to this property.

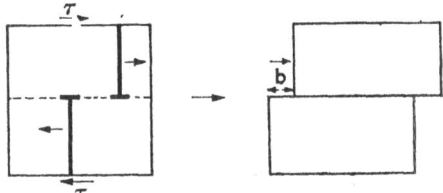

Figure 3.7. Shear of a crystal cube due to the passage of two dislocations of opposite signs

Again, in a crystalline material the Burgers vector AA' must be equal to a definite lattice spacing.

It should be noted that the Burgers vector lies in the direction of the dislocation line OO', while in the case of an edge dislocation (*Figure 3.4*) it is perpendicular to the dislocation line. This distinction will be of particular significance in the consideration of the spread of slip by a dislocation loop, which will be considered below. As a preliminary we shall examine the behaviour of two edge dislocations of the same Burgers vector lying on the same slip plane, when a shear stress is applied to the latter. As can be seen from *Figure 3.7*, if the shear stress is applied as indicated, the upper dislocation will move to the right and the lower one to the left, and the crystal will be sheared as shown; the effect is the same as would be obtained by the passage of a single dislocation right through the crystal. If however the sense of the shear stress were reversed the dislocations would move towards one another and, on meeting, a complete unfaulted plane of atoms would form, with the consequent disappearance of both dislocations. This phenomenon has its parallel with screw dislocations, and may be generalised in the rule that coalescence of two dislocations of the same type but opposite signs results in their mutual annihilation.

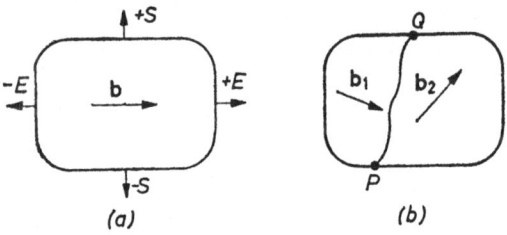

Figure 3.8. Dislocation loops (slip plane parallel to the surface of the paper): (a) all parts of the loop have the same Burgers vector, and the loop expands under an appropriate shear stress as indicated by the arrows at the edges; (b) a loop with different Burgers vectors in parts separated by the line PQ

Now, returning to the analogy of the runner carpet, we see that the ruck representing the dislocation line cannot terminate inside the carpet; it must stretch right across its entire width. In fact, the ruck may be regarded as the boundary line separating the slipped from the yet 'unslipped' part of the carpet. Similarly, a dislocation cannot terminate within a crystal. It must either terminate at the crystal walls, or it must exist in the form of a closed loop enclosing a slipped area, as shown in *Figure 3.8a*.

It is clear from the discussion of the characteristics of edge and screw dislocations that if the positive edge dislocation moves as indicated under a shear stress in the slip plane, the negative edge dislocation $- E$, as well as the screws $+ S$ and $- S$ will also move outwards, away from the centre of the loop, thus expanding the slipped area. That the Burgers vector of a closed loop must in fact be the same over its entire length can be shown by considering a loop (*Figure 3.8b*) divided by a line PQ into two parts with Burgers vectors b_1 and b_2 respectively. This implies that the area left of PQ has slipped by an amount differing from that by which the part to the right of PQ has slipped. The difference is in fact $b_1 - b_2$, so that PQ must be a dislocation having this Burgers vector. It follows that a dislocation loop, free from network nodes such as PQ, has the same Burgers vector throughout.

In general the corners of isolated loops will tend to be rounded, consisting of dislocations of the ' mixed ' type, which have partly edge and partly screw characteristics, behaving essentially as if they consisted of small segments of alternate edge and screw type. The question of the curvature of dislocations will be discussed in some detail later.

3.3 Stresses due to Dislocations

As can be seen from *Figure 3.5*, the crystal is compressed above the slip plane near the centre of a positive edge dislocation; below the slip plane it is in tension. The converse is of course true for a negative edge dislocation. Consequently the elastic, potential, energy stored in the lattice around two edge dislocations of the same Burgers vector but opposite signs shown for example, in *Figure 3.7*, could be reduced if the dislocations came close together, because the overlapping of compressed and dilated zones would diminish the local strains. In fact the potential energy gradient between the dislocations manifests itself as a mutual attraction; however, if the dislocations on the slip plane were of equal signs they would repel, somewhat like electric line charges of equal signs.

The state of stress due to dislocations in homogeneous, elastically isotropic, bodies can be evaluated by classical methods of elasticity. The results may be applied, with certain modifications, to crystals. Apart from the elastic anisotropy of most crystals, which we shall disregard, the calculations cannot be applied to a small volume contained within a cylinder of a few atomic spacings in diameter

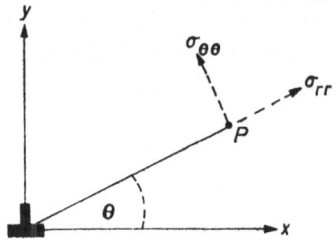

Figure 3.9. Stresses at a point in the crystal due to an edge dislocation lying along the z-axis

centred on the dislocation core, for in view of the large strains at the core, Hooke's Law, assumed in the calculations, is there no longer valid.

For an edge dislocation lying along the z-axis of a Cartesian co-ordinate system (*Figure 3.9*) the non-zero components of the stress tensor are

$$\sigma_{xx} = - D.y \frac{(3x^2 + y^2)}{(x^2 + y^2)^2} \qquad \sigma_{yy} = D.y \frac{(x^2 - y^2)}{(x^2 + y^2)^2}$$

$$\sigma_{zz} = \nu (\sigma_{xx} + \sigma_{yy}) \qquad \tau_{xy} = D.x \frac{(x^2 - y^2)}{(x^2 + y^2)^2} \tag{3.8}$$

where $D = Gb/2\pi(1 - \nu)$, G being the shear modulus, b the Burgers vector, and ν Poisson's ratio. In cylindrical co-ordinates one obtains

$$\left.\begin{array}{l} \sigma_{rr} = - (D/r) \sin \theta \\ \sigma_{\theta\theta} = - (D/r) \sin \theta \\ \tau_{r\theta} = (D/r) \cos \theta \end{array}\right\} \tag{3.9}$$

with σ_{zz} equal to ν ($\sigma_{rr} + \sigma_{\theta\theta}$). For a screw dislocation lying along the z-axis one obtains, as the only non-zero component of the stress,

$$\tau_{\theta z} = Gb/2\pi r \qquad (3.10)$$

which is independent of θ.

3.4 Tilt and Twist Boundaries in Crystals

As a consequence of their stress fields dislocations will exert forces on one another, and will tend to rearrange themselves in such a manner that the elastic energy of the crystal is reduced. Loss of dislocations by mutual annihilation may result, as well as the formation of relatively stable dislocation arrangements.

Referring to *Figure 3.9*, for example, edge dislocations of the same sign as the one at the origin will be attracted towards the y-axis if their x-co-ordinates are less than their y-co-ordinates; for τ_{yx} is then negative (equation 3.8). Hence, edge dislocations lying above the line $x = y$, which makes an angle of 45° with the x-axis, will tend to be swept into alignment along the y-axis. Once they have arrived in this position they in turn will ' rake in ' positive dislocations located close to them and above a similar line inclined by 45° to the axis, at the same time repelling nearby edge dislocations of negative signs. This discrimination and segregation process is therefore autocatalytic, and a dislocation wall or boundary may form. If the separation between dislocations in the wall is large compared with the lattice spacing, b, the parts of the crystal on opposite sides of the wall become tilted with respect to one another by an angle $\alpha = b/d$ (*Figure 3.10a*). In a bent crystal in which the process has been allowed to reach completion, so that only the excess dislocations of one sign necessary to maintain the bent shape remain, the dislocations will be arranged in tilt boundaries, approximately at right-angles to the slip planes, as indicated in *Figure 3.10b*.

If the radius of curvature R of the bent crystal is assumed to be large compared with the crystal thickness, and the dislocation walls are taken to be spaced a distance u apart

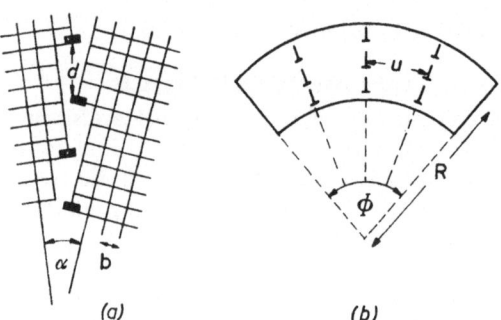

Figure 3.10. Structure of: (a) a tilt boundary;
(b) a bent, polygonised, crystal

then, as each wall contributes a tilt b/d to the angle ϕ

$$(R\phi/u)b/d = \phi \qquad (3.11)$$

The density of dislocations, given by the number intersecting unit area of crystal perpendicular to the glide plane, is

$$\rho = 1/ud$$

so that

$$\rho = 1/bR \qquad (3.12)$$

This equation enables one therefore to estimate the density of excess dislocations of one sign introduced into a crystal in the process of bending. On account of the polygonal shape assumed by the crystal as a result of this process, it is often referred to as ' polygonisation '. *Plate I* shows the formation of ' cell ' walls or ' sub-boundaries ' by a process of polygonisation in a grain of deformed polycrystalline

56

0°01 mm

Plate I. Glide polygonisation in 99·999 per cent pure polycrystalline magnesium after deformation at 200° K

magnesium. The glide planes intersect the surface in the slip lines running approximately diagonally from the right lower corner; the almost vertical sub-boundaries appear to have collected dislocations from the slip planes.

More complex mutual displacements of parts of crystals can be described by other arrangements of dislocations. Thus in *Figure 3.11* part of a crystal below the glide plane *ABCD* is held stationary, while the upper part is displaced by slip due to the introduction of a set of screw dislocations

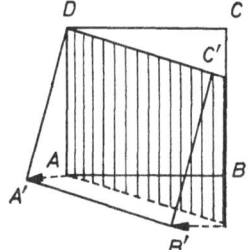

Figure 3.11. Formation of a twist boundary between parts ABCD and A'B'C'D of a crystal

parallel to *AD*. It becomes lozenge-shaped in the process. If now a second set of screw dislocations is introduced, this time parallel to *AB*, the upper part of the crystal regains its previous form. The deformation with the final position of the upper part at *A'B'C'D*, is equivalent to displacing the two parts of the crystal on opposite sides of the glide plane by twisting about an axis perpendicular to the slip plane. The resulting 'twist' boundary consists of a crossed grid of screw dislocations.

3.5 The Energy of Formation of Dislocations

The energy required to form an edge dislocation in a homogeneous, isotropic, elastic material may be obtained by evaluating the work necessary to displace the two surfaces formed by cutting a cylinder of radius r_1, and unit length, by an amount b, as indicated in *Figure 3.12*. We shall consider the energy stored in the entire volume of the cylinder except for a narrow core region of radius r_0 centred on the dislocation, where the displacement will be significantly less than b. This core will be discussed separately.

Now the shear stress on the slip plane at a point P a distance r from the origin will rise from zero at the time when the cylinder edges at A and A' coincide, to its full value $\tau = Gb/2\pi (1 - \nu)r$ (equation 3.9) when the dislocation is already situated at O. The mean stress in the course of introducing the dislocation will therefore be $\frac{1}{2}\tau$, and the

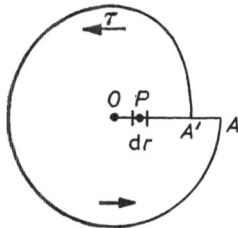

Figure 3.12. Formation of an edge dislocation in a cylinder of radius r_1 and unit length

work required to effect the displacement b over a strip of unit length and width dr is

$$dW = \tfrac{1}{2} [Gb^2/2\pi (1 - \nu)] \, dr/r$$

so that the total work to form a dislocation of unit length is

$$W = [Gb^2/4\pi (1 - \nu)] \ln (r_1/r_0) \qquad (3.13)$$

The energy of formation per segment of length b is

$$w = bW. \qquad (3.14)$$

If the core radius is taken to be about $4b$, and the range of the stress field of a dislocation is assumed to be rather less than the mean distance between dislocations in moderately deformed crystals, so that $r_1 \approx 1000 \, b$, one obtains

$$W \approx \tfrac{1}{2}Gb^2 \qquad (3.15)$$

This estimate is not very sensitive to variations in the ratio r_1/r_0, which appears in equation 3.13 only as a logarithm.

An upper limit to the energy stored in the core can be obtained by considering it to be so heavily distorted that its structure resembles that of a liquid. The energy within it cannot then exceed the latent heat of melting within a tube of molten crystal having the same volume as the core. On evaluating this one finds this energy to be negligible compared with W, which may therefore be regarded as the total energy of formation per unit length of edge dislocation. By analogy with the two-dimensional ' surface tension ', W is sometimes referred to as the ' line tension ' or ' line energy ' of the dislocation. The above treatment may be applied to screw dislocations; similar results are obtained.

Now in copper, for example, w (equation 3.14) is found to be about $2\frac{1}{2}$ eV, and since a dislocation must be many lattice spacings long to form a loop or arc enclosing a slipped area, the energy of formation of such loop or arc would be so high that generation of dislocations by thermally activated processes is impossible. Exceptions may occur under rather special conditions, for example in atomically thin wedges of metal, bounding holes produced by electropolishing. In such a case the least length of dislocation required to span the wedge near its edge may be only a few interatomic spacings, and in view of the thinness of the wedge the elastic energy stored within it may be very considerably less than that given by equation 3.13.

Dislocation-free crystals of germanium, silicon and other elements have been made by special methods; in the case of metals they can at present be produced only in the form of extremely fine whiskers. In general, dislocations are invariably present in crystals, and they can act in a number of ways as sources of new dislocations. One of these mechanisms we shall now consider.

3.6 The Frank – Read Source

The force F acting on unit length of dislocation due to a shear stress τ in the slip plane can be found by considering the movement of an edge dislocation a distance dr in the direction of its Burgers vector **b**. The slip displacement over this area is b, and the force $\tau . dr$; one therefore obtains for the energy expended:

$$F . dr = (\tau . dr)b$$

or

$$F = \tau b \qquad (3.16)$$

The same result would be obtained for screw dislocations.

We are now in a position to examine the effect of a shear stress in the slip plane on the shape of a dislocation. In *Figure 3.13*, let AB represent a short segment of length x of a dislocation firmly pinned at its two ends, e.g. at walls

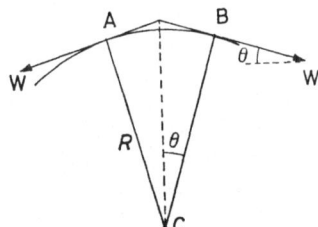

Figure 3.13. Equilibrium of a dislocation pinned at its ends, under a shear stress τ

of the crystal. Because of its line tension the dislocation will bow out in the slip plane under the applied shear stress until it attains an equilibrium position. This occurs when the force $(\tau b) x$ due to the stress is balanced by the opposing component of the line tension $2W\sin \theta$.

With θ taken small, also noting that $x = 2R\theta$, the condition of equilibrium then yields

$$\tau b = W/R \qquad (3.17)$$

and substituting for W from equation 3.15, the radius of curvature is found to be inversely proportional to the stress, and given by

$$R = \tfrac{1}{2}Gb/\tau \qquad (3.18)$$

Thus in the absence of a stress the dislocation will be

stretched straight. The radius of curvature will diminish as the stress is increased; consecutive stages are indicated by 0, 1 and 2 in *Figure 3.14*.

When the dislocation has become semicircular and the stress is increased further it is geometrically impossible to satisfy equation 3.18, and the dislocation becomes unstable, first developing re-entrant lobes, then breaking up into two parts due to annihilation of parts of the coalescing lobes of opposite sign which, assuming the dislocation was initially of the edge type, are screw segments, as shown in the

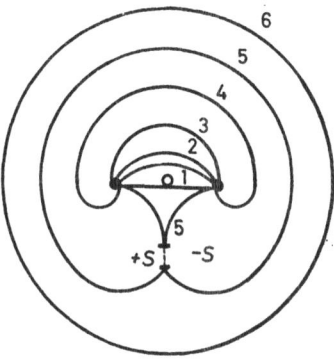

Figure 3.14. Consecutive stages in the expansion of a dislocation pinned at two points, under the action of an increasing shear stress

diagram. The pretzel-shaped part, numbered 5, straightens to become a dislocation loop, while the remaining, cusp-shaped, dislocation reverts to 1, thus recreating the source. The mechanism, first suggested by F. C. Frank and W. T. Read, is therefore capable of operating again, throwing off more loops. Various obstacles, such as 'forest' dislocations intersecting the slip plane containing the source, may interfere with it and thus arrest the 'mill' after a few loops have been generated. Evidence of the existence of Frank–Read sources in crystals has been obtained, though other mechanisms leading to an increase in the dislocation density can be imagined, and are probably of more frequent

61

occurrence. The pinning points may again be dislocations on slip planes intersecting the glide plane of the source dislocation, though impurities and other lattice defects could also provide effective local locks.

A crystal containing dislocations pinned at points spaced approximately l apart would therefore show a fairly abrupt onset of plasticity when R (equation 3.18) becomes equal to $\frac{1}{2}l$, which is the criterion for activating the sources. The flow stress of the crystal would therefore be given by

$$\tau = Gb/l \qquad (3.19)$$

In well-annealed metal crystals the free length of dislocations may be several microns (*Plate II*). If, for example, one takes $l = 5$ μm, equation 3.19 yields for copper a flow stress of about 25 kg/cm², which is close to the experimentally determined value. On comparing equations 3.5 and 3.19 one observes that the flow stress of a real crystal is by a factor of approximately $10b/l$ smaller than that of dislocation-free ideal crystals.

3.7 Preferred Slip Systems and Miller Indices

Some reference to slip systems has already been made in section 2.2, without however a consideration of the crystallographic nomenclature used to describe lines and planes in crystals. We shall now discuss the specification of lines and planes in crystallographic terminology.

Now, it is obviously desirable to use a reference system of axes bearing some relation to the crystal morphology. *Figure 3.15* shows such a set for a triclinic crystal where the lengths of edges of the unit cell a, b and c and the angles between the faces are all unequal. The units in which the axes are measured are multiples of the lattice lengths. The shaded plane, for example, is specified by (111), and the same set of indices would be used to describe the shaded plane in the face-centred cubic crystal shown in *Figure 2.1*.

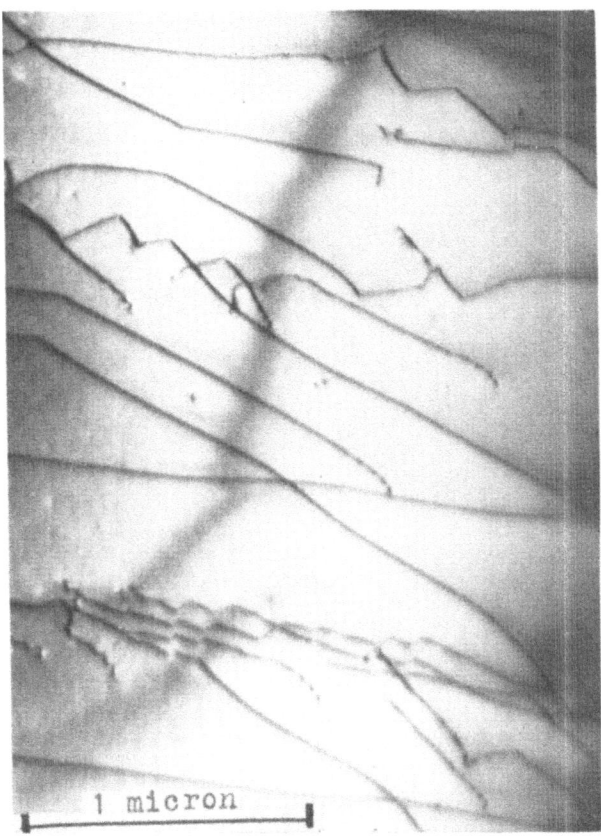

1 micron

Plate II. Electron transmission micrograph of dislocations in a slightly deformed copper foil less than ½ μm thick. The slip plane containing the dislocations is inclined at a few degrees to the plane of the foil, and some dislocations are seen to terminate at points where they meet one of the foil surfaces

The set of numbers in parentheses are known as the Miller indices of the plane, and for any given plane they are obtained as follows.   Consider a plane which, to give a specific example, makes intercepts $1a$, $0\cdot5b$ and $3c$ on the co-ordinate axes.   We now take the set of inverses of these numbers, i.e. $1/1$, $1/0\cdot5$ and $1/3$ and multiply all of them by the smallest number which will convert all the fractions to units.   In the present case multiplication by 3 yields (3, 6, 1), . which are the required Miller indices.   Similarly, the shaded plane in the sodium chloride crystal (*Figure 2.1b*), which

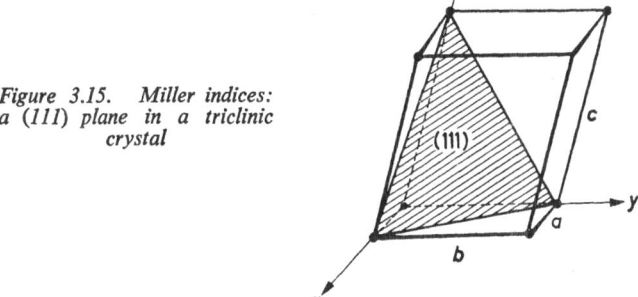

Figure 3.15.   Miller indices: a (111) plane in a triclinic crystal

intersects the x, y, z axes at 1, ∞, 1, has Miller indices (1, 0, 1).   If the intercepts on the axes are negative a minus sign is placed above the number, e.g. (2, $\bar{3}$, 0).

In order to specify the direction of a line it is necessary to give the co-ordinates of two points.   However the convention is to assume that the line passes through the origin, so that only one point has to be given explicitly.   Again, using a definite example, a line passing through the point with co-ordinates $1a$, $0\cdot5b$ and $3c$ would be described by [2, 1, 6], the numbers being proportional to the set 1, $0\cdot5$, 3, but without fraction or common factor.   As can be seen, a square bracket is used. Minus signs are again placed above the numbers; the convention is the same as with the planes.

The Miller index notation is particularly useful in the case of cubic crystals; the units of length, a, b and c are then equal. The Miller index of a plane (h, k, l) is then the same as that of the normal to the plane, $[h, k, l.]$ The angle θ between two planes is equal to that between their respective normals, i.e. between the lines $[h_1, k_1, l_1]$ and $[h_2, k_2, l_2]$, and is given by the equation

$$\cos \theta = \frac{h_1 h_2 + k_1 k_2 + l_1 l_2}{(h_1^2 + k_1^2 + l_1^2)^{\frac{1}{2}}(h_2^2 + k_2^2 + l_2^2)^{\frac{1}{2}}} \qquad (3.20)$$

It is readily checked, for example, that the planes (101) and ($\bar{1}$01) are mutually perpendicular, for the numerator in equation 3.20 is zero. For the angle between [001] and [101] one has $\cos \theta = 1/2^{\frac{1}{2}}$, and hence $\theta = 45°$.

Equation 3.20 is also useful if it is desired to establish whether a given line, for example [11$\bar{2}$], is parallel to a certain plane, e.g. (111). The scalar product [11$\bar{2}$].[111] given by the numerator in equation 3.20 must then be zero, as is in fact found to be the case.

The preferred slip systems in the face-centred cubic metals, which include Al, Ag, Au, Cu, Ni, Pb, Pd, Pt, Rh and others, are {111}, < 110 >, which are also operative in the cubic structure of the tetravalent homopolar crystals of diamond, germanium and silicon. In alkaline halides the slip planes and directions are {110}, < 110 > as indicated in *Figure 2.1*. The slip planes of body-centred cubic metals such as iron and tungsten are not unique, {112} and {110} may be operative; the slip direction remains however < 111 >. The braces and angular brackets indicate planes and lines of certain types respectively. Thus {112} is the shorthand notation for (112), (211), (121), (1$\bar{1}$2), etc.

DISLOCATION KINETICS AND LATTICE DEFECTS

4.1 Close Packing and Partial Dislocations

Structures typical of most metals can be regarded as consisting of closely packed spherical ions in contact. Only two crystal structures can be obtained in this manner, as can be seen by considering possible ways of stacking close-packed layers of the type shown in *Figure 4.1*. The numbered arrows indicate possible positions at which ions of adjacent sheets could be centred. If, for example the stacking sequence is 1, 2, 3, 1, 2, 3, 1 . . . the resulting lattice is face-centred cubic, as is apparent from *Figure 4.1b*, while if the stacking is such that only two positions are utilised, i.e. 1, 2, 1, 2, 1 . . . the close-packed hexagonal structure is obtained. Hexagonal metals are less common than face-centred cubic ones; they include among others beryllium, magnesium, zinc and cadmium.

In the course of plastic deformation the close-packed planes slip over one another without disturbing the stacking sequence. Thus an ion located at a certain ' 2 ' position in *Figure 4.1a* would be transferred to a crystallographically equivalent adjacent ' 2 ' position by the passage of a dislocation. Possible Burgers vectors must therefore join equivalent sites. However, a sheet of ' 2 ' ions could be imagined to slip over a ' 1 ' sheet by a sequence in which the transfer of ions from positions *A* to *B* (*Figure 4.1a*) occurred via ' 3 ' positions, as indicated by the Burgers vectors $\mathbf{b}_1$ and $\mathbf{b}_2$ into which $\mathbf{b}$ may be considered to have been resolved. While the vector sum of $\mathbf{b}_1$ and $\mathbf{b}_2$ is equal to $\mathbf{b}$, it is clear that

$$\mathbf{b}_1^2 + \mathbf{b}_2^2 < \mathbf{b}^2 \tag{4.1}$$

since the angle between b_1 and b_2 is 120°. Assuming elastic isotropy, it follows from equation 3.15 that replacement of the dislocation with Burgers vector b by two 'partial'

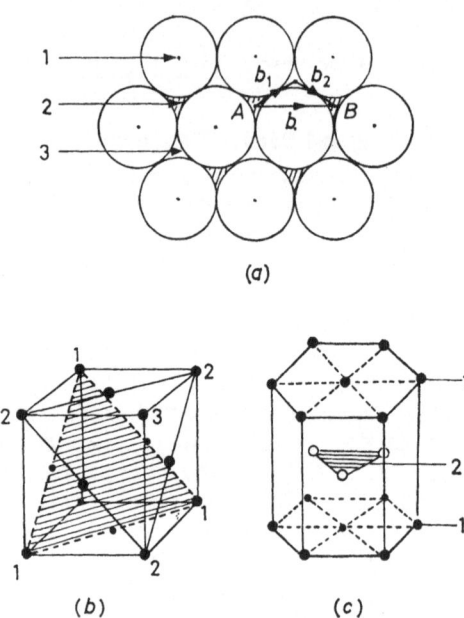

(a)

(b)   (c)

Figure 4.1. Close-packed structures. In (a) centres of a further layer of close-packed spheres could be placed either into positions 2 or into positions 3. The Burgers vector b can be split vectorially into $b_1 + b_2$. The face-centred cubic and hexagonal close-packed structures are obtained by stacking close-packed sheets in the sequences 1, 2, 3, 1, 2, 3, 1 . . . and 1, 2, 1, 2, 1 . . . respectively

dislocations with Burgers vectors b_1 and b_2 respectively would result in a reduction of the elastic energy of the crystal. The splitting of a glide dislocation into two partials is therefore energetically favoured here; that it does in fact occur is experimentally well documented. As energy would be required to recombine two partials, a force of repulsion must exist between them. However, they can move apart only over a definite, fixed distance, for in the process of separation a stacking fault is formed between them, and the available energy becomes expended. The stacking fault tends to draw the dislocations together, somewhat like a soap film possessing surface energy; the balance of attractive and repulsive forces acting on the partials will determine the equilibrium separation between

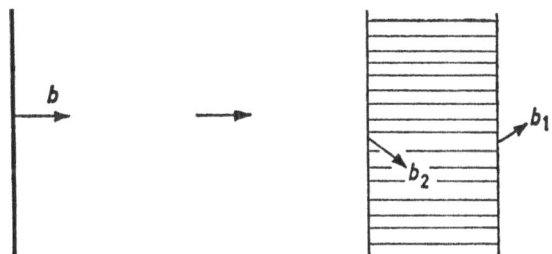

*Figure 4.2. Splitting of an edge dislocation of Burgers vector **b** into two parallel partial dislocations of mixed character and Burgers vectors **b**₁ and **b**₂. The shaded region represents the stacking fault*

them. Typical values of the width of the stacking-fault ribbon are about $1b$ for aluminium and $10b$ for copper, but larger values may occur in certain alloys such as stainless steel, in graphite and other crystals. That a stacking fault must form between the two partials (*Figure 4.2*) may be seen by considering that the passage of the first partial dislocation over a certain area of the slip plane of a face-centred cubic metal, for example, will change the stacking

sequence of 1, 2, 3, 1, 2, 3 to 1, 2 | 3, 1 ↓ 3, 1 | 2, 3, which

contains a 3, 1, 3, 1 sandwich characteristic of hexagonal crystals. The second partial dislocation completes the slip process, thereby restoring correct stacking. Similarly, stacking faults having face-centred cubic structure can be introduced into hexagonal metals in this manner.

Splitting of dislocations into partials is of considerable significance in relation to certain features of the plastic behaviour of crystals. We see for example that screw dislocations which, as we know, have a radially symmetric stress field and do not 'carry' an excess plane of atoms, could cross-slip from one glide plane to an intersecting one, as indicated in *Figure 4.3*. In aluminium and in body-centred metals where glide dislocations are little or not at

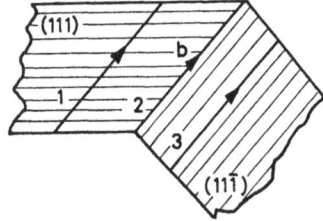

Figure 4.3. Cross-slip of a screw dislocation

all dissociated this process appears to take place relatively readily. Dissociated screw dislocations cannot cross-slip until they have recombined, at least over a certain length; hence rather special circumstances must prevail to facilitate this. Cross-slip is then more difficult and relatively rare.

4.2 Jogs and Point Defects

As can be seen from *Figure 4.4* slip on intersecting glide planes leads to the formation of steps on the dislocations, known as 'jogs'. The sharp bend in the dislocation at the jog distorts the lattice. A jog will therefore exercise a frictional drag on the dislocation containing it.

Now, remembering that the Burgers vector of a dislocation is the same at all of its parts, hence also at the jog, the latter can be seen (*Figure 4.5*) always to have edge character. In the case of the edge dislocation, denoted by E, the glide plane of the jog contains its Burgers vector, so that the jog can move with the dislocation in the direction *AB* without great difficulty. As will be seen, however, it cannot move at right-angles to the direction of *AB* without ploughing up the lattice and forming vacant lattice sites or interstitial ions in the process.

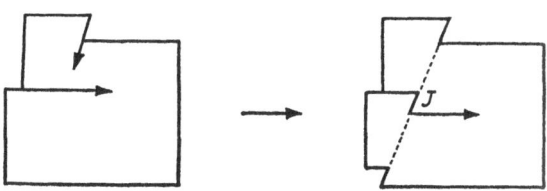

Figure 4.4. Formation of jogs on dislocations through inter-section by dislocations moving on another slip system

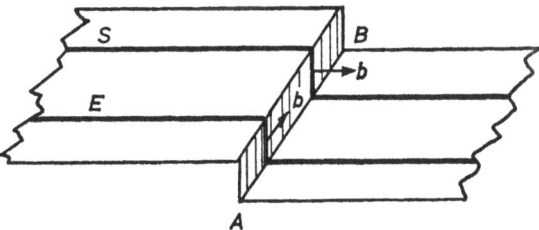

Figure 4.5. Jogs in edge and screw dislocations having Burgers vectors along and at right-angles to the direction of motion AB respectively. The jog in the screw dislocation has edge character and can move with the dislocation along AB only non-conservatively

By contrast, the Burgers vector of the jog in the screw dislocation *S* is perpendicular to the direction of movement *AB*. The jog could therefore migrate along the dislocation but not with the dislocation line, along *AB*, for in that case it would be forced to migrate at right-angles to its Burgers vector and, again, it would generate vacancies or inter-stitials. The mode of migration of a jog or dislocation

69

along its slip plane is termed 'conservative'; movement resulting in the formation of point defects is 'non-conservative'.

How point defects form through jog movement can be seen by considering the non-conservative (a) upward and (b) downward displacement of an edge dislocation in a monatomic sheet of crystal, illustrated in *Figure 4.6*; the slip plane is horizontal in both cases. In (a) the ion located at the lattice point '1' jumps away into an interstitial position '2', and the dislocation consequently climbs to

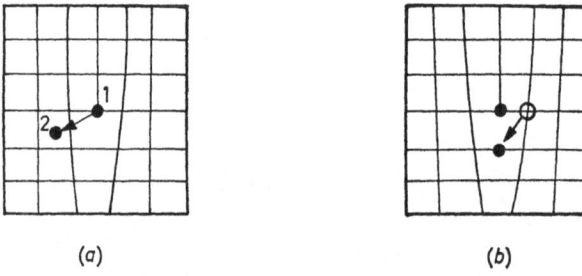

(a)   (b)

Figure 4.6.   Formation of (a) an interstitial and (b) a vacancy by non-conservative ' climb ' of an edge dislocation

the slip plane above the original one; in (b) an ion jumps below the centre of the dislocation, extending the latter and leaving a vacancy adjacent to it. The force on the jog arises from the stress acting on the dislocation containing it, so that the stress in the slip plane of the glide dislocation provides the energy required to form chains of point defects as the jog is being dragged along non-conservatively. Thermal activation may assist this process at elevated temperatures. Dissociated jogs may behave in a more complex manner than undissociated ones; there is some evidence which suggests that conservative movement may be difficult or even impossible in such cases.

4.3 Obstacles to Glide

A dislocation moving on its slip plane will be intercepted by obstacles, primarily by the ' forest ' dislocations threading its glide plane. In order to be able to move over significant distances the dislocation must either cut through the forest dislocations, with the resulting formation of jogs in the forest and glide dislocations, or the stress must ' extrude ' it past the forest dislocations as shown in *Figure 4.7*. In the latter case, when the arcs between the obstacles *A* and *B* have become almost semicircular the segments of opposite signs will pinch off, in a manner similar to that described in

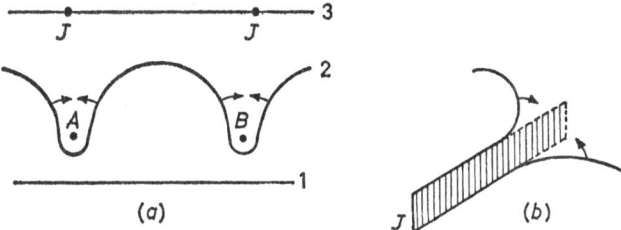

Figure 4.7. Obstacles to dislocation movement : (a) forest dislocations A and B; (b) a large jog, giving rise to a dipole

the discussion of the Frank–Read source. The dislocation, which may become jogged in the process, will then be able to move along under the applied shear stress. The jogged loops remaining threaded around the forest dislocations would contract under their own line tension and disappear, in certain cases leaving stable jogs in the forest dislocations. If a dislocation contains a ' large ' jog, i.e. a dislocation segment many interplanar spacings long (*Figure 4.7b*), an elongated dipole may be left in its wake.

Whether in the course of plastic deformation the forest is intersected as indicated in *Figure 4.7a*, or whether a more direct cutting of dislocations occurs may depend on the material and specific conditions, for example certain stress and temperature combinations. In the direct cutting process two jogs would form simultaneously with an expenditure of energy equal to an appreciable fraction of Gb^3 (equation 3.14); thermal activation could assist the process and a high temperature sensitivity of the flow stress would be expected. In the extrusion mechanism the work effecting ' extrusion ' is done by the applied stress, and no significant thermal activation would occur. The relatively low temperature sensitivity of the flow stress of annealed metal crystals suggests that an extrusion process occurs as a rule.

4.4 The Velocity of Jogged Dislocations

A dislocation moving in a crystal otherwise free from imperfections should be able to accelerate to a velocity equal to an appreciable fraction of that of sound in the material. In general, however, they move comparatively slowly, as has been found by direct observation by electron transmission micrography and by etch-pit methods. The relatively low mean velocities must be due to the periodic arrest of the dislocation at obstacles, to the drag of jogs and that of impurities which, by interacting with the stress field of the dislocation, tend to pin it, or to combinations of all three effects.

We shall consider the drag of jogs, for these are invariably formed if the material is plastically strained, irrespective of chemical purity. Vacancy generating jogs on screw dislocation will be taken as a specific illustration.

Now *Figure 4.8* shows a jogged screw dislocation with jogs spaced l_1, l_2, etc. apart, bowed out into arcs of radii of curvature R by an effective shear stress τ. The force on the jog J arises from the line tension W of the dislocation, and its component F perpendicular to the Burgers vector and along the direction of motion is

$$F = W(\sin\theta_1 + \sin\theta_2) \tag{4.2}$$

72

while the component perpendicular to F, tending to move the jog conservatively along the dislocation is

$$F' = W(\cos\theta_1 - \cos\theta_2) \qquad (4.3)$$

In general therefore the jog will move sideways as well as forward, the respective velocity components depending on F and F', and on the magnitudes of the energies required to effect the displacements. In the direction of F the motion is non-conservative; a point defect has to be formed for each interatomic displacement in this direction. Along the dislocation the jog can move conservatively, i.e.

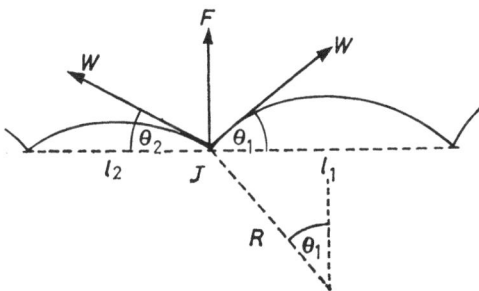

Figure 4.8. A screw dislocation dragging a jog J

by increasing l_1 at the expense of l_2, without migrating in the direction of F. Such a conservative displacement could not continue if the jog ran into parts of the dislocation not of purely screw character.

Although a jog may eventually run off the dislocation, out of the crystal, or become annihilated by encountering a jog of opposite sign, dislocations will acquire new ones in the course of their passage through the forest. At any given instant a dislocation will therefore be subjected, in general, to a drag arising from non-conservative jog displacements. We shall therefore examine the drag of an array of ' non-conservative ' jogs on a screw dislocation, assuming the jogs to be a fixed distance apart; this static distribution will be taken to be equivalent to the actual, dynamic one, where jogs are both formed and lost. As these 'equivalent' jogs move with the dislocation in the direction F their velocity and that of the dislocation must be the same. Now, if the energy of formation of a point defect is Q_0 then, since the work done by the line tension of the dislocation in the process of forming the point defect is Fb, where we take b to be the lattice spacing in the direction of F, the frequency ν_1 of forward jumps will be given by the Boltzmann relation

$$\nu_1 = \nu_0 \exp[-\,(Q_0 - Fb)/kT] \qquad (4.4)$$

for a thermally activated rate process in which the energy barrier Q_0 is reduced by the work contributed to point defect formation by the stress. Here ν_0 is an atomic vibration frequency of the order of 10^{12} per second, k is Boltzmann's constant and T the temperature in °K. Jumps in the reverse direction will be impeded by the stress, and the frequency for backward jumps will be

$$\nu_2 = \nu_0 \exp[-\,(Q_0 + Fb)/kT] \qquad (4.5)$$

The net forward jump rate will therefore be

$$\nu = \nu_0 \left[\exp\left(-\frac{Q_0 - Fb}{kT}\right) - \exp\left(-\frac{Q_0 + Fb}{kT}\right) \right] \qquad (4.6)$$

The second term is generally negligible compared with the first one; if, however, both are taken into account we can write equation 4.6 in the form

$$\nu = 2\nu_0 \exp(-\,Q_0/kT)\,\sinh(Fb/kT) \qquad (4.7)$$

Since the jog, and hence the dislocation, moves a distance b in the direction F in each successful jump, the dislocation velocity is

$$u = \nu b = 2\nu_0 b \exp(-\,Q_0/kT)\,\sinh(Fb/kT) \qquad (4.8)$$

and this is linear in F if $Fb \ll kT$.

The shear stress τ can be introduced into this equation as follows. On writing $l_1 = l_2 = \ldots = l_j$, where l_j is the equivalent or ' effective ' jog spacing, one obtains, noting that then $\theta_1 = \theta_2 = \theta$ (*Figure 4.8*)

$$2W\sin\theta = F; \quad R\sin\theta = \tfrac{1}{2}l_j \tag{4.9}$$

which yields with equations 3.15 and 3.18:

$$F = \tau b l_j \tag{4.10}$$

so that

$$Fb/kT = \tau b^2 l_j/kT \tag{4.11}$$

The product $b^2 l_j$ is known as the ' activation volume '.

WORK-HARDENING, RECOVERY AND CREEP

5.1 The Stress-Strain Curve of Single Crystals

We consider a cube-shaped single crystal of length of side L_0 containing one dislocation of Burgers vector **b**, as in *Figure 3.5*. When the dislocation has passed through the entire crystal the shear strain will be b/L_0. However, if it moves into the crystal from the left over a distance $x < L_0$, the shear strain will amount to only $(b/L_0)(x/L_0)$. When n dislocations have moved along parallel glide planes into the crystal the shear strain would be $\frac{1}{2}nb/L_0$, taking the mean value of x/L_0 for a uniform distribution of dislocations to be $\frac{1}{2}$. The ratio n/L_0^2 is the density ρ of glide dislocations, so that the shear strain due to them is

$$\gamma = \tfrac{1}{2}\,\rho\, bL_0 \qquad (5.1)$$

Provided that the number ρ' of forest dislocations threading unit area of the active glide planes under consideration remains constant, the mean distance between points at which forest dislocations intersect an active slip plane will be fixed and equal to $(1/\rho')^{\frac{1}{2}}$. According to equation 3.19 the flow stress will therefore remain constant, independent of the strain, the distance between pinning points on the glide dislocations being determined by ρ'.

Constancy of the density of forest dislocations can be maintained in the early stages of deformation of single crystals of many metals which have been well annealed. The deformation then proceeds without significant work-hardening up to strains depending on the perfection and dimensions of the crystal and its orientation with respect to the tensile axis, factors which may influence the onset of glide on intersecting slip systems. This 'easy glide', generally referred to as 'stage 1' is then followed by 'stage 2' of intense work-hardening (*Figure 5.1*); here interaction of dislocations on intersecting slip systems leads to the formation of stable dislocation networks.

As the stress is increased new dislocations move into the crystal, preferentially in regions where they are least impeded by the stress fields of existing ones. 'Soft' spots in the

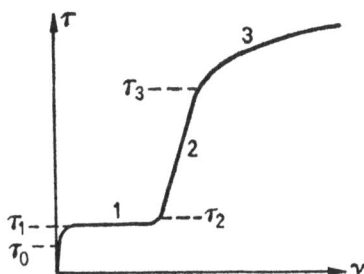

Figure 5.1. The work-hardening curve characteristic of single crystals. All three stages may not always be realisable

crystal, containing a 'below average' density of dislocations therefore become less soft as the deformation proceeds, and the network tends to assume similar mesh dimensions and geometrical features throughout the crystal. The mesh diminishes with increasing stress, without however significant changes in its shape. Groups of large loops as shown in *Figure 5.2* may therefore become superseded by sets of smaller ones as a result of the overlapping and interaction of dislocations moving from the periphery of the large loop into its interior.

Now if the slip distance at any stage of this parcellation process is L, the corresponding mean spacing between

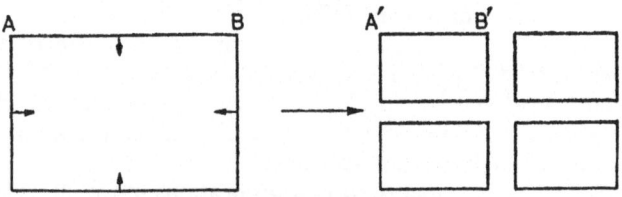

Figure 5.2. Parcellation of a large loop into a set of four geometrically similar ones. (Schematic)

adjacent dislocations is d, and the separation of active slip
planes D, then the maintenance of geometrical similarity
suggests the relation

$$L = c_1 D = c_2 d \tag{5.2}$$

or

$$L = (c_1 c_2 D d)^{\frac{1}{2}} = (c_1 c_2)^{\frac{1}{2}}/\rho^{\frac{1}{2}} \tag{5.3}$$

where ρ is the dislocation density and c_1 and c_2 are constants
determining the geometry of the dislocation distribution.

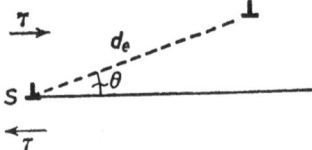

Figure 5.3. A disloca-
tion source S of edge
type in the field of
an edge dislocation
of the same sign,
also subjected to an
applied shear stress τ

The origin of the constancy of the ratios of the dimensions L, D, d, expressed by equation 5.2, may be illustrated
as follows. Consider a dislocation source, e.g. of edge type,
l_s long, subjected to an applied shear stress τ in its slip
plane. Let it also be subjected to an opposing shear stress
due to an edge dislocation of equal sign fixed in the lattice
at a distance d_e from the source, as shown in Figure 5.3;
the angle θ is assumed to be small to avoid mathematical

difficulties. The source will operate if the shear stress complies with the equation

$$\tau - \frac{Gb}{2\pi (1 - \nu)d_e} \approx \frac{Gb}{l_s} \tag{5.4}$$

which follows from equation 3.19, assuming $d_e \gg l_s$. If τ is the flow stress at which a large number of sources become activated simultaneously then one can infer from equation 5.4 that

$$\tau \approx \frac{Gb}{l_s} \left[1 + \frac{l_s}{2\pi (1 - \nu) d_e} \right] \tag{5.5}$$

and that the flow stress over a small range of source lengths would not vary significantly provided l_s/d_e remained approximately invariant. If one takes for the distribution of source lengths and corresponding spacings d_e

$$l_s = l_0 (1 + x), \qquad d_e = d_0 (1 - ax), \qquad | x | \ll 1$$

with

$$a = 2\pi (1 - \nu) d_0/l_0$$

then τ is independent of x, and the ratio l_s/d_e is very nearly constant. Conservation of geometrical network similarity is probably a consequence of the tendency of dislocation movement and formation to occur so as to introduce the least elastic strain energy into the crystal for any given plastic strain increment.

Now returning to *Figure 5.2*, we note that if the slip distance L is approximately equal to a characteristic dimension of the rectangles, such as AB and $A'B'$, then by equation 5.1 the shear strain would be

$$\gamma \approx \tfrac{1}{2} (1/Dd) bL \tag{5.6}$$

If the distance between forest dislocations, and hence between 'hard' immobile pinning points on any given dislocation is taken to be $(Dd)^{\frac{1}{2}}$ (equation 5.3), then one obtains from equation 3.19, irrespective of whether it is applied to a Frank–Read source or to the 'extrusion' mechanism considered in section 4.3, the relation

$$\tau = Gb\rho^{\frac{1}{2}} \tag{5.7}$$

The product Dd has here been equated to $1/\rho$. Equations 5.3, 5.6 and 5.7 yield the linear work-hardening law

$$\tau/\gamma = h; \qquad h = G/\tfrac{1}{2}(c_1\,c_2)^{\frac{1}{2}} \tag{5.8}$$

characteristic of ' stage 2 '.

The values of the constants of proportionality c_1 and c_2 cannot be estimated without further detailed analysis. However, with reasonable values, for example $L = 1$ mm, $D = 10$ μm and $d = 1$ μm, corresponding to a dislocation density of 10^7 cm^{-2}, one obtains $c_1 = 100$ and $c_2 = 1000$, so that

$$h \approx G/150 \tag{5.9}$$

which is in good agreement with measured values, available mainly for close-packed metal crystals. The shear modulus varies only slowly with temperature, and h would therefore be expected to be rather temperature-insensitive; this again is confirmed by experiment.

When a certain dislocation density has been attained, characterised by the temperature-dependent shear stress τ_3 (*Figure 5.1*), the dislocations begin to interact sufficiently strongly to be able to rearrange themselves by glide poly-gonisation and cross-slip, thereby reducing the strain energy stored in the crystal. Mutual annihilation may also occur. These processes will take place already in the course of deformation above τ_3, and the rate of accumula-tion of dislocations will therefore be less than if such ' dynamic recovery ' were absent. The strain will now be due not only to slip by the dislocations present in the crystal; it will include contributions from dislocations which are no longer in the material. One may therefore write instead of equation 5.8

$$\gamma = \frac{\tau}{h}[1 + a\,(\gamma - \gamma_3)] \qquad (5.10)$$

where a is a constant which will in general increase with increasing temperature and decrease with increasing strain rate. The term $\gamma - \gamma_3$ is taken to be zero at strains below γ_3; it represents a contribution to the total strain by recovery processes. If one rewrites equation 5.10 in the form

$$\gamma = \tau/h' \qquad (5.11)$$

with

$$h' = h/[1 + a\,(\gamma - \gamma_3)] \qquad (5.12)$$

the coefficient of work-hardening h' in ' stage 3 ' of the work-hardening curve is seen to decrease uniformly to zero with increasing strain. The observed rapid decrease of the slope of the work-hardening curve as deformation proceeds in stage 3 is qualitatively in agreement with equation 5.12.

5.2 Work-hardening of Polycrystals
Easy glide is not observed in polycrystals, since operation of one glide system only would not allow all the grains to deform simultaneously and uniformly. This is apparent, for example from equation 2.4 which shows that the tensile flow stress

$$\sigma = \tau_{0r}/\cos\,\theta \cos\,\phi$$

would have to be exceedingly high for certain crystals, or grains, which would therefore fracture rather than deform, unless alternate more favourable slip systems were available. Hexagonal metals for example, in which the close-packed basal plane is the preferred glide plane, with other systems less easily induced to propagate slip, are appreciably less ductile than face-centred cubic crystals which have several non-parallel slip systems of the same type (section 3.7) to accommodate glide.

Although more than one slip system must operate in polycrystalline hexagonal metals to allow them to attain the observed, not insignificant, plastic strains without fracture, the basal planes are generally most active, with slip on other systems occurring mainly near grain boundaries. Stage 2 of rapid work-hardening, which involves dislocation interaction on intersecting systems, does not therefore seem to take place to an appreciable extent in hexagonal metal polycrystals, although it is observed in single crystals.

'Linear' hardening, corresponding to stage 2 in single crystals, is readily demonstrated in face-centred cubic metals provided they are sufficiently well annealed to have a tensile flow stress less than σ_3, which corresponds to τ_3 in single crystals. The magnitude of the coefficient of work-hardening $d\sigma/d\epsilon$ can be deduced from h (equation 5.9) as follows.

Consider a polycrystalline tensile specimen consisting of grains with a mean flow stress in shear τ. On deforming the specimen plastically by a tensile strain increment $d\epsilon$ the amount of work done per unit volume is $\sigma.d\epsilon$, and this must be equal to $\tau.d\gamma$, where $d\gamma$ is the mean 'equivalent' shear strain increment in the crystals. On equating these energies:

$$\sigma.d\epsilon = \tau.d\gamma$$

one obtains

$$\sigma/\tau = d\gamma/d\epsilon \tag{5.13}$$

Now the mean shear stress would be expected to be somewhat less than the maximum shear stress $\frac{1}{2}\sigma$; in fact it may be shown that

$$\tau \approx \tfrac{1}{3}\sigma \tag{5.14}$$

which, substituted into equation 5.13 yields

$$\gamma \approx 3\epsilon \tag{5.15}$$

Thus from equations 5.8, 5.14 and 5.15 one obtains,

$$d\sigma/d\epsilon = H \qquad (5.16)$$

where

$$H \approx 9h \qquad (5.17)$$

and the same equations could be used to adapt equation 5.10 for polycrystals. In view of the intracrystalline processes visualised in the derivation of equation 5.10 the modification derivable from it for the stress–strain relation of polycrystals would not be expected to have grain size as a variable. In fact, in face-centred cubic metals the stress–strain relation is relatively insensitive to grain dimensions. In hexagonal metals, however, the slip distance L_0 introduced into equation 5.1 should be equal to a substantial fraction of the grain diameter, for slip on non-basal planes would be relatively ineffective in preventing the passage of dislocations across an entire grain. We therefore use equation 5.1, appropriate to such a model of glide in the grains, together with equation 5.7, and obtain the ' parabolic ' work-hardening law

$$\tau^2/\gamma = 2G^2b/L_0 \, (1 + \alpha) \qquad (5.18)$$

where the factor $(1 + \alpha)$ has again been introduced to allow for the contribution to the strain by dislocations effectively removed by recovery processes. It should be noted that the dislocation density in equations 5.1 and 5.7 is the average density in the metal; individual grains will have different densities $\rho(L)$, depending in general on the grain dimension. If the shear strain is not to fluctuate unduly from grain to grain, so that coherency is maintained, equation 5.1 suggests that the product $\rho(L)$. L would be roughly constant. Thus the dislocation density would then tend to be below average in large grains and above average in small grains. There is experimental evidence in support of this conclusion.

On using equations 5.14 and 5.15 in equation 5.18, one obtains

$$\sigma^2/\epsilon = \chi \qquad (5.19)$$

with the temperature-dependent coefficient of ' parabolic ' work-hardening equal to

$$\chi(T) = 54G^2b/L_0(1 + a) \qquad (5.20)$$

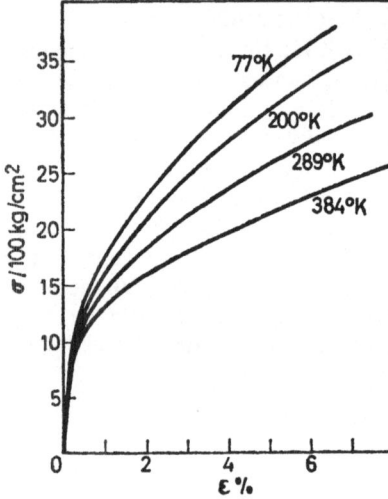

Figure 5.4. Tensile stress–strain curve of polycrystalline cobalt deformed at a strain rate of about 0·1 per cent per second

One may infer from equations 5.19 and 5.20 that if a series of well annealed polycrystalline specimens of the same material but different grain dimensions L_0 are subjected to tensile tests, then the flow stresses measured at the same strain will be proportional to $\chi^{\frac{1}{2}}$, and hence to $L_0^{-\frac{1}{2}}$. Such a proportionality, sometimes referred to as Petch's relation, has been studied extensively in steels. It explains the well-known opinion that fine-grained polycrystals are generally stronger than coarse-grained ones.

Figure 5.4 shows the tensile stress-strain curves of annealed polycrystalline cobalt at various temperatures,

84

well within the range where the metal is close-packed hexagonal. The applicability of equation 5.19 is apparent from *Figure 5.5* in which the ' plastic ' part of the curves,

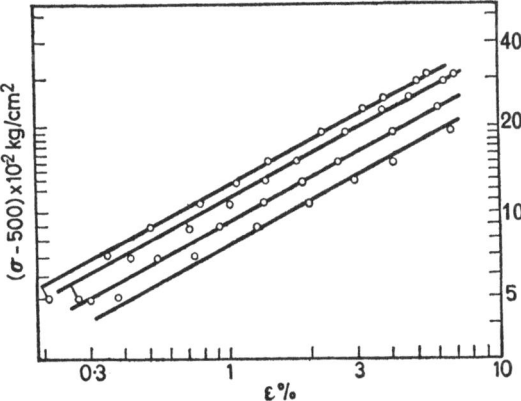

Figure 5.5. Log–log representation of the work-hardening curves shown in Figure 5.4. The slopes of all lines are $\frac{1}{2}$

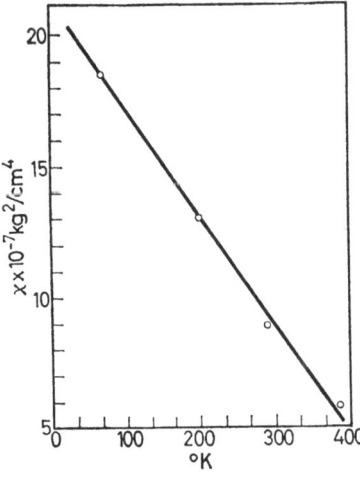

Figure 5.6. Temperature dependence of the coefficient of 'parabolic' work-hardening in polycrystalline cobalt. Measured points were obtained from Figure 5.5

i.e. above $\sigma = 500$ kg/cm^2, have been plotted on log-log co-ordinates; the lines drawn through the points all have slopes of $\frac{1}{2}$. The effect of temperature on χ is shown in *Figure 5.6*. If the slip distance in the metal is assumed to be about half the grain diameter, then in this case $L_0 = 55$ µm. With $G = 9 \cdot 1 \times 10^5$ kg/cm^2 and $b = 2 \cdot 51$ Å one obtains for temperatures close to $T = 0°$K, i.e. with $a = 0$ in equation 5.20, $\chi = 20 \times 10^7$ kg/cm^2, which agrees well with values obtained experimentally at low temperatures.

5.3 Stress Relaxation

If a rod made of metal, most plastics or other ductile materials, is subjected to a tensile test and the test is suddenly stopped, with the material held at the final strain attained, the stress would be observed to decrease in the course of time, eventually attaining a new, stable, value. This phenomenon, known as stress relaxation at constant strain, can be understood in terms of residual plastic deformation, resulting in the smoothing out of the internal stress peaks in the material. The elastic energy of the local stress peaks is then partly dissipated by ' plastic ' work, without however resulting in an overall extension of the rod.

Provided the process occurs at temperatures sufficiently low to prevent significant contributions to atom transport by diffusion mechanisms, the internal stress peaks become reduced as a consequence of ' microscopic ' local shears. This ' dynamic ' recovery must be distinguished from high-temperature recovery and the associated high-temperature creep in which thermally activated processes contribute to significant structural changes in the material undergoing recovery.

We shall examine this relaxation in more detail, taking as a specific example a crystalline material, e.g. a metal. To simplify the calculation we shall assume that a single crystal cube of side length L has been subjected to a shear

86

strain γ_{tot} at which it is then held. Since this total strain consists of the elastic and plastic components,

$$\gamma_{tot} = \gamma_{el} + \gamma \qquad (5.21)$$

the constancy of the total shear strain implies

$$d\gamma = - d\gamma_{el} = - d\tau/G \qquad (5.22)$$

where G is the shear modulus and τ the applied stress, which is, of course, a function of time. A plastic strain increment $d\gamma$ arises from the displacement of ρL^2 glide dislocations by a small distance dx along their respective slip planes so that, as in section 5.1

$$d\gamma = (\rho L^2) (b/L) (dx/L) = \rho.b.dx \qquad (5.23)$$

where we have considered the temperature to be low enough to enable us to assume that the density ρ of moving dislocations does not vary in the course of the relaxation experiment. From equation 5.23 one finds, directly,

$$d\gamma/dt = \rho.b.u \qquad (5.24)$$

where u is the instantaneous velocity of the dislocations.

On using equation 5.22 and the relation for the dislocation velocity (equations 4.8 and 4.11) the rate of stress relaxation is found to be given by

$$- \frac{d\tau}{dt} = G\rho b^2 v_0.\exp\left[- (Q_0 - \tau b^2 l_j)/kT\right] \qquad (5.25)$$

the ' high stress ' form of equation 4.8 being used. The mean effective stress acting on the glide dislocations, appearing in the exponential term, is assumed equal to the applied shear stress; this is reasonable, for l_j is also taken to be a mean value. A more convenient way of writing equation 5.25 is

$$- \frac{d\tau}{dt} = A.\exp\left[- \frac{Q_0}{kT}\left(1 - \frac{\tau}{\tau_0}\right) \right] \qquad (5.26)$$

where

$$\tau_0 = Q_0/b^2 l_j \qquad (5.27)$$

and

$$A(\rho) = G\rho b^2 \nu_0 \qquad (5.28)$$

The physical significance of τ_0 may be appreciated by considering that stress relaxation could not occur at the observed, non-zero, rates at low temperatures unless as $T \to 0$ also $\tau \to \tau_0$ (equation 5.26). With our definition of τ the stress τ_0 should therefore be equal to the flow stress in shear at very low temperatures.

The dislocation density is assumed to be determined by the stress $\tau(0)$ at the time $t = 0$ at which the relaxation under constant strain begins; A is time invariant during relaxation at a given strain.

Integration of equation 5.26 yields

$$(kT/Q_0\tau_0) \exp[-(Q_0/kT)(\tau/\tau_0)] = A(t+t_0) \exp - (Q_0/kT)$$

where t_0 is a constant of integration. On taking logarithms of both sides and differentiating with respect to $\ln(t + t_0)$ one obtains

$$d\tau/d \ln(t + t_0) = -\tau_0 (kT/Q_0)$$

which yields the equation of 'logarithmic' stress relaxation

$$\tau(0) - \tau = r \ln(t + t_0) \qquad (5.29)$$

with

$$r = \tau_0 kT/Q_0 \qquad (5.30)$$

It is frequently possible to neglect t_0 after the first few seconds of relaxation, and the evaluation of r is then particularly simple. Equation 5.30 may be applied to tensile stress relaxation by replacing τ_0 by $\sigma_0/3$, as in equation 5.14. Remembering that $\sigma(0) \to \sigma_0$ as $T \to 0°K$, equation .30 can be used to evaluate Q_0 from measurements of r and τ made at low temperatures.

Experimental work on stress relaxation on copper, cobalt, nickel, iron, magnesium, and uranium, as well as on polymers such as polyethylene, on cork and other materials, have shown that the logarithmic form of stress relaxation is of wide occurrence at low temperatures. The reason for this universality lies no doubt in the generality of the assumptions embodied in equation 5.22 and in the assumed linear stress dependence of the activation energy (equation 5.25), which would be valid, at least as a first approximation, for most thermally activated processes in which the stress reduces the energy barrier.

For polycrystalline metals deformed in the ' parabolic ' stage of work-hardening, values of Q_0 agree quite well with the activation energies for vacancy formation, suggesting that the non-conservative drag of intersection jogs is responsible for the frictional force on moving dislocations. The activation energy to form interstitial ions in general exceeds that for vacancy formation by a factor of about 3; interstitials do not in fact appear to form in appreciable concentrations in metals in the course of plastic deformation.

Further, equations 5.27 and 5.30 show that measurements of r provide a means for evaluating the average jog spacing l_j; this is generally found to lie in the range $100-1000b$. With the approximation $\tau = \tau_0$, valid at low temperatures, the relation for the flow stress (equation 3.19) yields with equation 5.27:

$$l/l_j \approx Gb^3/Q_0 \qquad (5.31)$$

and for most metals this ratio is approximately 5, in reasonable agreement with experiment.

5.4 Logarithmic Creep

The above discussion of stress relaxation has shown that the state of stress in a material is not determined only by the strain, but also by the temperature and the loading rate; the latter may of course be zero. One can express this formally by writing

$$d\tau = \frac{\partial\tau}{\partial t}dt + \frac{\partial\tau}{\partial T}dT + \frac{\partial\tau}{\partial\gamma}d\gamma \qquad (5.32)$$

G

In the case of isothermal tests, which we shall consider, the term $\partial\tau/\partial T$, which takes into account mainly changes of elastic constants with temperature, is zero. Further, if the material is subjected to a constant stress, $d\tau = 0$ and one obtains

$$\frac{\partial\tau}{\partial t}\,dt = -\,\frac{\partial\tau}{\partial\gamma}\,d\gamma$$

Hence the shear rate under constant stress is

$$\frac{d\gamma}{dt} = -\,[1/h(\gamma)]\,\frac{\partial\tau}{\partial t} \qquad (5.33)$$

where $h(\gamma)$ is the coefficient of work-hardening which would be measured at a strain γ, at a sufficiently fast shear rate to remain unaffected by time-dependent recovery effects. The partial differential $\partial\tau/\partial t$ represents the rate of stress relaxation at constant strain. Recovery effects at low temperatures are sufficiently small to enable us to assume that $h(\gamma)$ remains at its initial value if the strain is slightly increased; this approximation is equivalent to assuming that the slope of the stress strain curve is constant over a small range of strains close to γ. Then, if both sides of equation 5.33 are multiplied by $t + t_0$ and integrated, one finds that

$$\frac{d\gamma}{d\ln(t + t_0)} = -\,\frac{1}{h(\gamma)}\cdot\frac{\partial\tau}{\partial\ln(t + t_0)} \qquad (5.34)$$

or, on using equations 5.29 and 5.30:

$$\gamma - \gamma(0) = (\tau_0 kT/Q_0 h)\ln(t+t_0) \qquad (5.35)$$

which represents the equation of 'logarithmic' creep. It may again be adapted for use with tensile deformation by using equations 5.14 and 5.15.

As $h(\gamma)$ generally decreases with increasing values of τ_0 the slope of the γ versus $\ln(t + t_0)$ curves should increase with increasing values of $\gamma(0)$. This conclusion, and the

90

simple relation between logarithmic stress relaxation and creep, expressed by equations 5.34 and 5.35, are well borne out by experiment.

5.5 High-temperature Creep

If plastically deformed crystalline materials are held for some time at relatively high temperatures, generally in excess of about $0 \cdot 4$ T_m (°K), where T_m is the melting temperature, they are observed to soften and the dislocation density diminishes. In polycrystals grain growth may also occur, the larger grains tending to consume the smaller ones, growing at their expense. After heavy deformation, such as can be induced by rolling, recrystallisation may set in; new grains nucleate and grow into the old, deformed ones, the stored elastic energy providing the principal driving force. High-temperature recovery, frequently termed ' annealing ' is used extensively in metal forming processes. A wire, for example, can be softened in this manner after leaving one die before being subjected to further reduction in another.

However, in high-temperature applications of materials, where the emphasis is on load bearing capacity, the enhanced susceptibility to softening and deformation arising from thermal recovery may be undesirable, for example in components of steam and gas turbines, in boilers and equipment used in the generation of electric power, or in space vehicles. The relatively slow deformations occurring in materials at elevated temperatures even under constant stresses are known as ' high-temperature creep '.

Thermal recovery tends to soften the material, but as the flow stress is thereby reduced the constant applied stress will again deform it, inducing renewed hardening. The essential feature of this dynamic balance between softening and hardening in isothermal creep may again be represented, formally, by means of equation 5.32; as the stress is constant $d\tau = 0$. However, by contrast with logarithmic creep the coefficient of work-hardening must now be considered to depend not only on the strain but, as the structure of the material changes during creep, also on time.

We shall take the flow stress τ of the crystal, e.g. a metal, to be given by equation 3.19, l being the structure-sensitive parameter. This stress could be measured, in principle, if the material undergoing creep could be cooled to a temperature at which the deformed state is stable, without introducing structural changes during the cooling. By differentiating equation 3.19 one finds for the rate of relaxation at constant strain

$$\frac{\partial \tau}{\partial t} = - \frac{Gb}{l^2} \cdot \frac{\partial l}{\partial t} \qquad (5.36)$$

which, with equation 5.33 yields

$$\frac{\mathrm{d}\gamma}{\mathrm{d}t} = \frac{Gb}{hl^2} \frac{\partial l}{\partial t} \qquad (5.37)$$

The partial differential coefficient represents the time rate of change of the characteristic dimension of the dislocation mesh which would be observed if the creep were suddenly stopped at the strain γ by removing the stress. Equations 5.36 and 5.37 show that the creep rate is determined by the ratio of the instantaneous rates of recovery and work-hardening.

When the constant stress is first applied to the material, loading conditions are akin to those encountered in a rapid tensile test, and the strain attained after a fixed, short, interval may obey a work-hardening law, such as is given by equation 5.19 for example. Following this ' initial ' rapid extension the strain rate will diminish in the course of transient or ' first stage ' creep until a dynamic equilibrium is established in which hardening and recovery occur at constant rates (*Figure 5.7*).

The creep rate then remains almost steady, and no appreciable structural changes occur in the material in this

' second stage ' or ' equilibrium ' creep, except for the gradual accumulation of localised internal damage which eventually leads to fracture. The accelerated creep, or ' stage 3 ', often observed just prior to fracture, is indicative of a progressive loss of intergranular coherency.

A detailed analysis of the cybernetics of high-temperature creep processes is still outstanding. However, considerable insight into creep mechanisms can be obtained by considering certain features of equation 5.37, making the

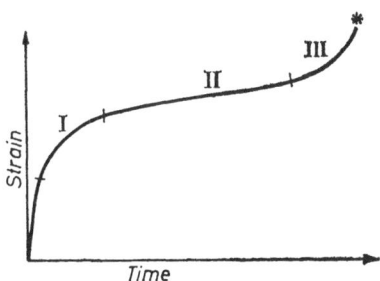

Figure 5.7. The three characteristic stages of high-temperature creep, following after the initial, rapid extension

assumption that variations in the dislocation mesh dimensions occur at a rate determined by the dislocation velocity, such that equation 5.37

$$\frac{\partial l}{\partial t} = cu \qquad (5.38)$$

In equilibrium creep, which we shall now consider, c will be taken to be a constant of order unity. One then obtains from equations 5.37, 5.38 and 4.8, using the high-stress approximation, the equation for the equilibrium tensile creep rate:

$$\frac{d\epsilon}{dt} = \frac{Gb^2 c v_0}{3hl^2} \cdot \exp\left[-(Q_0 - \tfrac{1}{3}\sigma b^2 l_j)/kT\right] \qquad (5.39)$$

where equations 5.14 and 5.15 were used to replace shear stresses and strains by equivalent tensile parameters. The values of h, l and l_j are assumed to be those characteristic of the equilibrium creep. If reasonable assumptions are made, for example $h = G/200$ and $l = 10^{-4}$ cm, also taking $c = 1$, the pre-exponential fraction in equation 5.39 is found to be about 4×10^5 per second for copper. This agrees quite well with measured values obtained at about 700°C, which is equivalent to about $0 \cdot 7 \; T_m$ (°K) in this metal. In this case, and generally above about $0 \cdot 5 \; T_m$, Q_0 is not identifiable with the activation energy for vacancy formation, as in logarithmic creep; it tends to be close to the activation energy of self-diffusion. This observation can be explained if it is considered that diffusional processes can occur readily at high temperatures, so that formation of a vacancy at a jog would not necessarily allow the jog to climb unless either the dislocation moves so fast that the jog is displaced from the vacancy before the latter can jump back to it, or if the vacancy, once formed, jumps away from the jog.

The first alternative is improbable in conventional high-temperature creep tests, as the dislocations would not move fast enough. In the second case the chain of events facilitating non-conservative jog displacements consists of the formation of a vacancy, followed by its diffusion away from the jog. The activation energy would then comprise the energies of vacancy formation and migration; their sum is in general close to the activation energy of self-diffusion.

5.6 Creep Fracture

Vacancies transported into the material by the dislocations in the course of creep will tend to segregate in the relatively open structure close to the grain boundaries or at local stress fields due to impurities in solid solution and, also, around inclusions. Such heterogeneous precipitation, leading to the aggregation of vacancies into small holes can be a source of fracture. If, for example, the damage attains a critical state in which small pores coalesce to form cracks the metal may fail rapidly. On the basis of the assumed accumulation of damage up to a critical amount a high creep rate would be expected to lead to a high rate of this process and hence to a relatively short creep life of the specimen, and vice versa. Also, since the second stage creep would in general prevail during most of the time during which the specimen is under load, one would expect to find the time to fracture t_f to be approximately inversely proportional to the equilibrium creep rate, i.e.

$$\left(\frac{d\epsilon}{dt} \right)_{II} \approx m/t_f \qquad (5.40)$$

where m is a dimensionless constant numerically close to the strain at fracture. The equation is reasonably well obeyed by pure metals and solid solutions, but not by highly heterogeneous materials.

In practice materials may be subjected to creep in use over many years; criteria of technical creep strength must then be specified in relation to the function and expected creep life of the material or structure in which it is embodied. For example, an alloy may be specified for a certain application such that under stated stresses and at a certain temperature it should not creep faster in tension than $0 \cdot 01$ per cent per hour when the creep strain has attained $0 \cdot 2$ per cent. Alternatively it may be required that after a fixed period under load at a given temperature, e.g. after 45 h, the specified stress should not lead to strains greater than $0 \cdot 2$ per cent, and the creep rate should not exceed $0 \cdot 01$ per cent per hour after, say, 30 h on test.

5.7 Point Defect Concentrations

In view of the importance of point defects, particularly vacancies and vacancy aggregates, in plasticity and fracture we shall attempt to obtain an estimate of the vacancy concentrations arising in the course of work-hardening. We shall again use face-centred cubic metals as our example, and make the simplifying assumption that all the vacancies are formed in rows by non-conservatively moving jogs, one vacancy being formed per displacement b of the dislocation.

According to equation 5.31 the spacing between non-conservatively moving jogs l_j may be expressed in terms of the dislocation mesh l by

$$l \approx 5 \, l_j \qquad (5.41)$$

which, on writing $l = \rho^{\frac{1}{2}}$, yields with equation 5.3:

$$L = 5 \, (c_1 \, c_2)^{\frac{1}{2}} \, l_j \qquad (5.42)$$

Now the number of jogs per unit length of dislocation will increase as the dislocation migrates over the slip distance L; they will first be relatively far apart, but the spacing will gradually diminish. The mean spacing between jogs on a given dislocation, considered over the period of its passage, will therefore be larger than the final value l_j referred to in equation 5.42, say $c_j \, l_j$, where we shall assume $1 < c_j < 10$. The total number of vacancies generated per centimetre of dislocation line in the course of slip will be

$$N = (L/b) \, (1/c_j l_j) \qquad (5.43)$$

and with the previous estimates of $c_1 = 100$ and $c_2 = 1000$, also taking $c_j = 5$, we obtain from equations 5.42 and 5.43:

$$N \approx 300/b \qquad (5.44)$$

For most metals this yields

$$N \approx 10^{10} \text{ cm}^{-1} \qquad (5.45)$$

Measurements of the excess resistivity due to point defects in deformed metals at low temperatures suggest that this estimate may be too high by a factor of about 10. It should be noted, however, that a uniform dispersion of vacancies is generally assumed in calculating vacancy concentrations from excess resistivity data, and this may lead to an under-estimate of N.

Thus a heavily cold-worked metal with a dislocation density of 10^{12} cm^{-2} should contain about 10^{21} vacancies per cm^3 due to cold-work. If these were condensed they would 'fill' a cube of about $2 \cdot 5$ mm length of side. Their presence should lead to a detectable reduction in the density of the material. Such density changes are observed, in part however they also arise from dislocations. On heating the metal to temperatures where vacancies become mobile, e.g. about 400°C in copper, self-diffusion should lead to a rapid drop of the vacancy concentration to the rather small equilibrium concentration; its magnitude at temperatures at which vacancies can diffuse at significant rates is given approximately by $\exp(-Q_0/kT)$, where Q_0 is the energy of formation of a vacancy. Evidence that vacancies also accumulate in ionic crystals on deformation at rates comparable to those here estimated has been obtained.

FRACTURE AND FATIGUE

6.1 Brittle Fracture

Two principal types of fracture are generally distinguished. Fracture preceded by a significant amount of plastic deformation is known as ' ductile ', otherwise it is ' brittle '. The latter, which we shall consider first, occurs when plastic flow is inhibited, be it by the effective locking of dislocations by precipitates or elements in solid solution or by the pre-existence or formation of cracks and imperfections acting as local stress raisers in the material. All materials can be embrittled if the temperature is lowered sufficiently. Glass, sealing wax, germanium, silicon and other materials, though ductile at temperatures close to their melting points, are brittle at ordinary temperatures; as was previously mentioned, rubber cooled in liquid nitrogen may be shattered with a hammer.

In most materials the brittle strength, defined as the maximum tensile stress withstood without the occurrence of brittle fracture, is low compared with the ideal strength the fault-free material would be expected to exhibit. The source of brittle fracture has therefore to be sought in the presence of structural defects.

Most of the theoretical work on brittle fracture has in fact been developed on the basis of the ' crack ' concept introduced by A. A. Griffith in 1924. He considered that brittle materials contained microscopic cracks which could extend their lengths abruptly if the tensile stress across them exceeded a definite critical value. This spread of cracks was assumed to lead to brittle fracture.

As long ago as 1913, C. E. Inglis had calculated the stress distribution in an elastic plate containing an elliptic crack of length $2c$, subjected to a tensile stress σ (*Figure 6.1*). He found that the tensile stress is increased at the ends of the crack by the ' notch factor ' to $2(c/R)^{\frac{1}{2}}\sigma$, R being the radius of curvature at the ends of the long axis of the ellipse. The elastic energy stored in the plate due to the presence of the crack

$$W(\sigma) = \pi c^2(\sigma^2/E) \tag{6.1}$$

increases with the crack length and the applied stress. Griffith considered that the crack would spread by extending

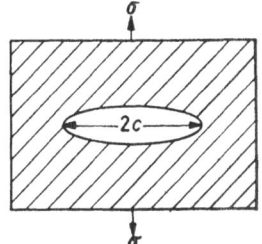

Figure 6.1. An elliptical Griffith crack subjected to a tensile stress σ

its length, and hence the fracture surfaces, when the stress attained a critical value σ_B, the brittle strength. Its value depended on the specific surface energy s of the crack faces, the latter requiring an energy

$$W(s) = 4cs \tag{6.2}$$

for their formation. The crack would spread when $\partial W(\sigma)/\partial c$ just exceeded $\partial W(s)/\partial c$, and this criterion yields

$$\sigma_B = (2sE/\pi c)^{\frac{1}{2}} \tag{6.3}$$

An estimate of the theoretical cohesive strength can be obtained from equation 6.3 on replacing c by the smallest physically meaningful length, an interatomic spacing in the preferred cleavage plane of the material. In freshly drawn fine glass rods σ_B has been found to be close to the theoretical value, but it diminished by a factor of 30 or more in the course of several hours, suggesting that cracks exceeding 1000 interatomic spacings in length must have formed in that time. Direct evidence in support of this inference was obtained by microscopy of the surface. In brittle metals the relation

$$\sigma_B \propto L_0^{-\frac{1}{2}}$$

is observed, where L_0 may be the mean grain radius or diameter. This suggests that the damaging cracks initially extend across the grains or run along large areas of certain grain boundaries.

The weakening effect of stress raisers, such as notches on the surface of the material, or the presence of sharp inclusions within it, are well known. A classical example is provided by the internal notches due to graphite flakes in cast-irons. The flakes embrittle the irons in tension; in structural applications cast-irons are therefore more usefully employed under compressive loads. Their brittle strength and toughness can, however, be increased appreciably if the graphite is allowed to form in spheroidal rather than flaky form. This can be achieved by alloying the melt, for example with magnesium; the resulting product is known as 'SG' (spheroidal graphite) iron.

6.2 Ductile Fracture

In single crystals deforming by slip on one preferred slip system, fracture may occur through 'slipping off'; one part of the crystal shears away entirely from another. Otherwise ductile fracture can occur as a result of the spread of small fissures, developing in 'bands' of closely spaced slip planes which have been particularly active in the deformation. The fissures form at points of high dislocation density where 'debris' left by dislocation interaction, and intense short-range stresses at local dislocation complexes, facilitate decohesion. Cracks may also form at inclusions and other flaws in the course of deformation.

The spreading of cracks from the surface to the interior of stressed thin metal foils has been observed by electron transmission microscopy. Propagation is generally ahead of the tip, where the stresses are highest due to the notch effect. A profusion of dislocations may be seen moving into or away from the tip, transporting material or ' vacancies'. Plastic deformation thus clearly facilitates extension of the damage.

The damage eventually leading to fracture appears to accumulate in the course of deformation. In copper for example the length of individual cracks and the number of cracks increase approximately linearly with deformation, but coalescence of cracks in the late stages of deformation does lead to a rather higher growth rate of the most ' dangerous ' cracks. One would therefore expect a relation of the form

$$c = c_0 (1 + a\epsilon^{1+\beta}) \tag{6.4}$$

to describe the crack growth kinetics reasonably well, $2c_0$ being the crack length in the unstrained material, and a and β positive constants appropriate under conditions of tensile deformation. If the stress across the cracks is taken to be proportional to the applied tensile stress σ then, writing $\sigma = \sigma_f$ at fracture, equation 6.3 becomes

$$\sigma_B = C\sigma_f = (2sE/\pi c_f)^{\frac{1}{2}} \tag{6.5}$$

where C is a constant of proportionality and c_f refers to the half length of the cracks at fracture. From equations 6.4 and 6.5 one obtains

$$\sigma_f^2 = K/(1 + a\epsilon^{1+\beta}) \tag{6.6}$$

where K is a constant which would not be expected to be significantly temperature dependent. If, as in the case of the cobalt specimen (*Figure 5.4*) the work-hardening can be described very well by the relation

$$\sigma^2 = \chi(T) . \epsilon \tag{6.7}$$

which also holds when $\sigma = \sigma_f$, then by eliminating the strain between equations 6.6 and 6.7, one finds that

$$\sigma_f^2 = K/[1 + a (\sigma_f^2/\chi)^{1+\beta}] \tag{6.8}$$

Now, as c_f would be expected to be much larger than c_0 (equation 6.4), the term $a(\sigma_f^2/\chi)^{1+\beta}$ should also be

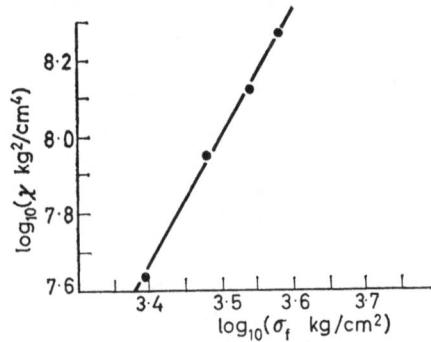

Figure 6.2. Log–log representation of the χ versus σ_f relation for cobalt. The slope of the curve yields $m = 0 \cdot 30$

appreciably larger than 1, so that one should have approximately

$$\sigma_f^2 \propto (\chi/\sigma_f^2)^{1+\beta} \tag{6.9}$$

or where $m = (1 + \beta)/(4 + 2\beta)$

$$\sigma_f \propto [\chi(T)]^m \tag{6.10}$$

The limiting values of m are therefore $\frac{1}{4}$ and $\frac{1}{2}$, i.e. for $\beta = 0$ or $\beta \to \infty$. *Figure 6.2* shows that the hypotheses made in deriving this result appear to be reasonable; the value obtained for cobalt is $0 \cdot 30$, corresponding to $\beta = 1$. For pure copper deformed in the same temperature range one obtains $m = 0\cdot40$, which is also well within the permitted limits of m. Although the uniqueness of the preceding model has not been established, the results are consistent with the view that fissures and the simultaneous increase of the applied stress eventually lead to the rapid spreading of cracks and consequent fracture.

If the specimen were deformed in shear, for example by twisting a rod, wire or tube, an equation corresponding to

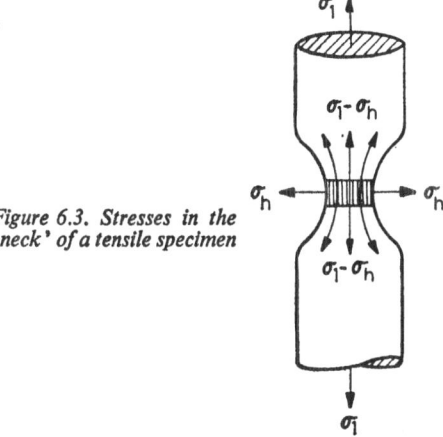

Figure 6.3. Stresses in the 'neck' of a tensile specimen

6.4 would still apply, but in the absence of applied tensile stresses the local internal tensile stresses would be expected to be relatively small, and large shear strains might be anticipated to precede failure. This is in agreement with experience; shear strains of 300 per cent or more are readily obtained by twisting metal wires such as silver or aluminium at ordinary temperatures, and these strains are greatly in excess of the corresponding values of $3\epsilon_f$ (equation 5.15), where ϵ_f is the strain at fracture in tensile specimens of the same material.

Also, if high hydrostatic stresses are superimposed upon the applied stress in a tensile test ductility is found to be enhanced. This can be explained if cracks are inhibited from opening up by the hydrostatic compression. Conversely, ductility is in general lowered if the specimen is notched. The origin of this effect can be seen on considering a tensile specimen which has a ' neck ', the latter being equivalent to a circumferential notch (*Figure 6.3*). The forces due to the applied stress σ_1 give rise to a transverse component in the neck region, so that the stress system can be resolved into an axial component $\sigma_1 - \sigma_h$ and a hydrostatic tension σ_h. If the flow stress in tension of a notch-free specimen is σ then for the necked sample

$$\sigma = \sigma_1 - \sigma_h$$

Consequently σ_1 must be increased to $\sigma + \sigma_h$ before flow will set in around the neck. The hydrostatic component of the stress will assist in the opening up of cracks, so that the material is embrittled at the same time as the ' apparent ' flow stress σ_1 is increased. In plastic materials σ_h cannot be increased indefinitely, e.g. by making the notch very sharp; it may be shown that σ_1 can attain a value of at most 3σ.

6.3 Fatigue

Fatigue is a mode of fracture which occurs when materials are subjected to alternating loads over prolonged periods at stress levels which would not lead to failure under static loading. It is a universal phenomenon, observed in most solids, and a common specific fracture mechanism cannot therefore exist, except in so far as cyclic loading leads to a continuous accumulation of damage which, as in the case of static fracture, eventually results in rupture.

Although the problem was studied by Albert almost 130 years ago, general awareness of the fact that the tensile strength is not a measure of the resistance of materials to fracture under all conditions encountered in practice was aroused only early in the 20th century as a consequence of the mysterious failure of some cast-iron bridges.

The essential features of fatigue, and measures which have to be taken to prevent its incidence, are nowadays quite well known, yet despite such powerful methods of investigation as electron transmission microscopy and X-ray micro-focus techniques it is likely to remain a quarry for research for some time to come.

Fatigue is a serious problem with metals for these are most widely used in dynamically loaded structures and machines. We shall therefore confine our attention to

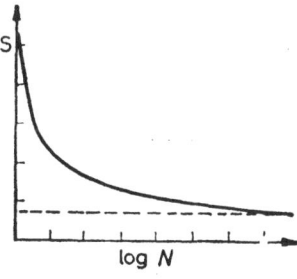

Figure 6.4. The relation between the peak stress S and the number N of cycles to fracture in alternating tension–compression with zero mean stress

metals and alloys, without thereby making a significant sacrifice of generality.

When a metal tensile specimen is subjected to tension–compression cycles with zero mean stress one finds in general that the relation between the peak stress S and the number of cycles to failure is of the form shown in *Figure 6.4*, known as the S–N or Wöhler curve. Semi-logarithmic co-ordinates are used. In practice the curve rarely becomes truly horizontal, so that there is no ' safe range ' of stresses, and fracture would ensue after a sufficiently large number of stress cycles. In certain alloys which have a well defined yield point, for example in plain carbon steels, the curve may have an asymptote, shown as a dashed line in the figure.

The stress corresponding to its intersection with the *S*-axis is then the ' true ' fatigue limit.

Determining the true fatigue limit is unnecessary and uneconomical in practical applications, and a ' technical ' fatigue limit is determined instead. This may be specified as the stress at which the material will endure a given number of cycles, for example 10 million, without fracture. The number of cycles leading to fracture at a given stress is often referred to as the ' fatigue strength ' or ' endurance ' at that stress.

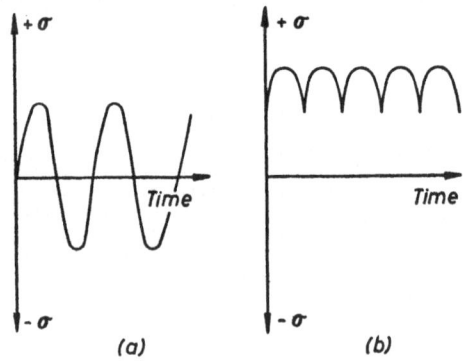

*Figure 6.5. Fatigue cycles with (a) zero and (b)
positive mean stress*

If the loading is not simple harmonic, with zero mean stress, but a constant tensile or compressive stress is super-imposed upon the oscillatory component, then a positive or negative mean stress results (*Figure 6.5*). For a given peak stress symmetric stressing, with zero mean stress, will in general lead to a longer fatigue life than the use of asym-metric cycles with the same peak stress. This is mainly due to the more complete healing of fatigue damage in the first case, for microscopic local stress peaks developed in, say, a positive cycle will be more nearly annulled by a complete reversal of the applied stress than by a partial reversal only. The $S-N$ curve for asymmetric stressing will therefore lie below that for simple harmonic loading.

In the early stages of fatigue testing, specimens will generally evolve an appreciable amount of heat due to plastic work, and they may work-harden somewhat. Later fissures develop and spread from slip bands at the surface, eventually leading to fracture. The surface is a preferential seat of damage initiation, for slip is less confined there than in the interior, and corrosive effects may also assist in the degradation of the structure at the surface.

The fatigue strength of metals can often be enhanced by treatments which render the surface more resistant to deformation; fracture then tends to start at the interface between the hard surface layer and the softer core. Stress raisers, such as sharp notches, corners, key-ways, rivet holes and scratches can lead to an appreciable lowering of the fatigue strength of metal components. Good surface finish and corrosion protection are frequently desirable to enhance fatigue resistance. Fatigue is essentially a low-temperature problem; at temperatures relatively high with respect to the melting point, fracture (and hence specimen life) are governed by creep.

The mechanism of metal fatigue is complex, and not fully understood. Available evidence shows, however, that damage accumulates gradually, as in ductile fracture and creep, until the specimen or component is weakened to the point where ordinary processes of fracture terminate its useful life. This view finds support in the observation that fatigue life is not significantly affected by the frequency of the loading cycles over wide ranges of frequency, but depends mainly on the total number of stress reversals.

Formation of cracks and cavities in the course of plastic deformation is favoured by extensive interaction between dislocations. In fact, in pure face-centred cubic metals the true fatigue limit seems to coincide with the stress at which the third, parabolic, stage of work-hardening begins. This stage, as was pointed out before, is accompanied by pronounced dynamic recovery resulting from dislocation interactions.

The gradual accumulation of damage is also clearly indicated by the observation that if a specimen is annealed after it has spent part of its expected fatigue life, the latter cannot in general be extended, unless the specimen is allowed to recrystallise by annealing at temperatures relatively high with respect to the melting point. A further similarity between fatigue and fracture lies in the temperature dependence of the technical fatigue limit and the fracture strength, which seems to be approximately the same in both cases. In commercial alloys the technical fatigue limit generally lies between $0 \cdot 3$ and $0 \cdot 5$ of the ultimate tensile stress. This range can however serve only as an approximate guide, for with materials of relatively complex structures it is difficult to make simple, precise, generalisations.

Fracture surfaces of fatigued metals generally show a smooth, lustrous, region, due to the polishing and abrading effects arising from attrition at fissures. The remaining parts of the fracture surface, over which failure occurred through weakening of the specimen by the reduction of its load bearing cross-section by surface cracks and fissures, may look duller and coarser, as it is essentially due to static fracture.

Incipient fatigue can sometimes be detected, even though fissures are not yet detectable, by a pronounced increase in the damping capacity or ' internal friction ' of specimens cut from the fatigued piece. This enhancement of the damping capacity is largely a consequence of the dissipation of part of the elastic energy of the vibrating or oscillating specimen by friction at crack interfaces. The method is however destructive; it is not possible in most cases to cut test specimens from operating structures or machines. It is therefore only useful as a research tool. A more detailed discussion of the principles involved in the study of internal friction will be given in the next chapter.

RHEOLOGICAL MODELS AND DAMPING CAPACITY

7.1 Damping Capacity

Dissipation of elastic energy, which manifests itself, for example, in the gradual decay of oscillations of a torsion pendulum and of the vibration amplitude of a tuning fork, even if air friction is practically eliminated altogether, is invariably observed in all materials. Energy may be dissipated in the formation of, or through abrasion at, cracks, by the displacement of ions from their mean positions in the lattice due to the to-and-fro movement of dislocations, by work expended in pulling dislocations away from point defect or impurity pinning points, and by a variety of other thermal, electrical, magnetic and diffusion processes; all have the common feature however of absorbing some of the elastic energy of oscillation or vibration stored in the material.

The simplest model of a damped solid consists of an ideally 'Hookean' element such as a longitudinally vibrating elastic rod, coupled in parallel with a viscous element or 'body'; a spring and a dashpot are in general used as the respective symbols, as shown in *Figure 7.1*.

For tensile deformation Young's modulus E and Trouton's coefficient of viscous traction ζ are used, while for deformations by shear they are replaced by the shear modulus G and the coefficient of viscosity η. With a Newtonian liquid in the dashpot ζ and η are stress and time independent. On taking σ_ζ to be the component of the

110

applied tensile stress σ acting on the dashpot only, one has

$$\sigma_\zeta = \zeta \cdot d\epsilon/dt \qquad (7.1)$$

where $d\epsilon/dt$ is the tensile strain rate of both spring and dashpot. The parallel-connected combination of spring and dashpot shown in *Figure 7.1* is known as the Voigt or Kelvin body. It is a solid, for it will not suffer permanent

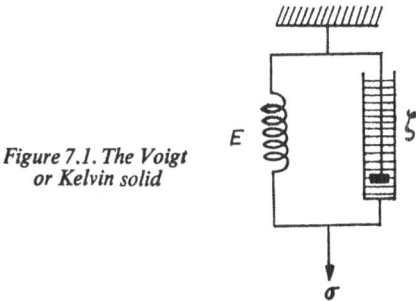

Figure 7.1. The Voigt
or Kelvin solid

deformation, always regaining its initial shape when the stress σ is removed. In the shear representation the equation equivalent to 7.1 is

$$\tau_\eta = \eta \cdot d\gamma/dt \qquad (7.2)$$

while the respective equations for the Hookean element are

$$\sigma_E = E \cdot \epsilon \qquad (7.3)$$

and

$$\tau_G = G.\gamma \qquad (7.4)$$

Again, as in equations 5.14 and 5.15, we may assume that σ_E is a constant multiple of τ_G, etc., and that, similarly, ϵ is proportional to γ. The relations

$$\frac{\sigma_\zeta}{\sigma_E} = \frac{\zeta}{E} \cdot \frac{d\epsilon/\epsilon}{dt}$$

$$\frac{\tau_\eta}{\tau_G} = \frac{\eta}{G} \cdot \frac{d\gamma/\gamma}{dt}$$

obtained from equations 7.1 to 7.4, then imply that

$$\zeta/E = \eta/G$$

and since, for an incompressible material, $E/G = 3$ (equation 1.26), we obtain the frequently quoted relation

$$\zeta = 3\eta \qquad (7.5)$$

between Trouton's coefficient of viscous traction and the coefficient of viscosity.

If one sets the body into oscillation, by releasing a mass which has first been suspended from it, one can evaluate the energy dissipated per cycle by the dashpot as follows. Assuming that the damping is sufficiently small to enable us to write for the strain over the period of one cycle

$$\epsilon = \epsilon_0 \cos \omega t \qquad (7.6)$$

where ϵ_0 is taken to be constant over that particular cycle, i.e. for $0 \leqslant t \leqslant 2\pi/\omega$, where ω is the frequency of oscillation in radians per unit of time, the energy dissipated is

$$\Delta w = \int_{t=0}^{2\pi/\omega} \sigma_\zeta \frac{d\epsilon}{dt} \ dt \qquad (7.7)$$

and this yields with equations 7.1 and 7.6:

$$\Delta w = \pi \zeta \omega \epsilon_0^2$$

With the assumed constancy of ϵ_0 over the cycle considered the elastic energy stored per unit volume is, by equation 1.20:

$$w = \tfrac{1}{2} E \epsilon_0^2$$

and one has

$$\Delta w/w = 2\pi (\zeta/E) \omega = 2\pi (\eta/G) \omega \qquad (7.8)$$

which is independent of ϵ_0. The ratio η/G is generally referred to as the relaxation time of the system.

Several measures of the damping capacity are in use. Of these the 'logarithmic decrement' and 'Q' are related to $\Delta w/w$ as shown in equation 7.9

$$\tfrac{1}{2}\,\Delta w/w = \pi/Q = \text{log. dec.} \tag{7.9}$$

The assumption that the viscosity is Newtonian, used in the above discussion of the Voigt–Kelvin model, is generally not restrictive in considerations of damping capacity, for the latter is in most cases measured at sufficiently small stresses to justify the assumption of a linear relation between stress and strain rate, implied by the use of Newtonian viscosity. It can be seen, for example, that the velocity u of dislocations, and hence the creep rate (equations 4.8 and 5.24) increase linearly with the stress if the latter is sufficiently small, and the dislocation density is constant, as could be the case at temperatures which are not too high in relation to the melting point. Under these conditions the flow of the material will therefore appear to be Newtonian.

7.2 The Standard Linear Solid

Solutions of polymers and other materials of high molecular weight often behave in a viscoelastic manner, i.e. like elastic solids when subjected to rapid loading, for example bouncing like a rubber ball, yet flowing like a viscous liquid, with permanent loss of shape, when subjected to static or slowly varying stresses. The simplest model with such behaviour is the Maxwell liquid, shown in *Figure 7.2*. It exemplifies creep as well as stress relaxation. The total strain ϵ now consists of the component $\epsilon_E = \sigma/E$ due to the elastic element, and a contribution made by the dashpot, defined by

$$\sigma = \zeta \,.\, \mathrm{d}\epsilon_\zeta/\mathrm{d}t$$

so that

$$\mathrm{d}\epsilon/\mathrm{d}t = \mathrm{d}\,(\epsilon_E + \epsilon_\zeta)/\mathrm{d}t$$

i.e.

$$\mathrm{d}\epsilon/\mathrm{d}t = E^{-1}\mathrm{d}\sigma/\mathrm{d}t + \sigma/\zeta \tag{7.10}$$

113

which is the differential equation describing the behaviour of the model.

If the stress is maintained constant, the first term on the right-hand side of equation 7.10 is zero, and the material flows, or creeps, at a constant tensile strain rate σ/ζ. At a constant strain ϵ_0, the strain rate $d\epsilon/dt$ is zero, and

$$d\sigma/dt = -\sigma (E/\zeta) \tag{7.11}$$

The stress then relaxes from its initial value in accordance with the exponential relation

$$\sigma(t) = \sigma(0) . \exp(- Et/\zeta) \tag{7.12}$$

It can be seen that after a period equal to the relaxation time ζ/E the stress has dropped to $1/e$ of its initial value.

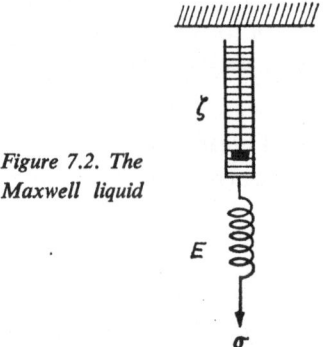

Figure 7.2. The Maxwell liquid

The simple rheological models so far discussed are rarely adequate to describe the behaviour of real materials even at low stresses, and more complex bodies consisting of combinations of Voigt–Kelvin and Maxwell bodies are then employed. A model which has been used extensively in the study of damping phenomena is the 'Standard Linear Solid' (*Figure 7.3*). It can account for the commonly made observation that the total reversible deformation attained by, say, a wire subjected to a constant stress does not occur in its entirety at the instant of loading; the wire continues to extend slightly at a diminishing rate for some time after the load has been applied. Neither the Voigt–Kelvin nor the Maxwell bodies can explain this behaviour adequately. This reversible but time-dependent extension, following the elastic deformation, has been termed by C. Zener ' anelastic ' to distinguish it from immediate, ideal, elastic response on the one hand and from irreversible, non-elastic, deformation on the other.

If the loading rate is so high that the plunger in the dash-pot has moved only very little by the time the full load is applied the dashpot may be regarded as rigid during that period, and by considering that the stresses acting on both spring elements give rise to the same strain in each, one finds that the effective ' unrelaxed ' elastic modulus is E. The subsequent movement of the dashpot under load eventually renders the spring with modulus ΔE ineffective, and the final, relaxed modulus is $E - \Delta E$. The increment ΔE is known as the ' modulus defect '.

By a procedure similar to that adopted above, one finds for the standard linear solid the equation

$$\sigma + a_\sigma \frac{d\sigma}{dt} = (E - \Delta E) \left(\epsilon + a_\epsilon \frac{d\epsilon}{dt} \right) \qquad (7.13)$$

where the relaxation time at constant strain a_σ and the retardation time at constants stress a_ϵ are given by

$$a_\sigma = \zeta/\Delta E; \quad a_\epsilon = [E/(E - \Delta E)] a_\sigma \qquad (7.14)$$

In practice $\Delta E \ll E$, so that the relaxation and retardation times will be numerically almost equal.

If the solid is subjected to a periodic stress $\sigma = \sigma_0 \cos \omega t$, one finds that the damping capacity is now given by

$$Q^{-1} = \frac{\Delta E}{E} \left(\frac{\omega a}{1 + \omega^2 a^2} \right) \qquad (7.15)$$

where the relaxation and retardation times have both been equated to a, i.e. E has been written for $E - \Delta E$. On plotting Q^{-1} as a function of the frequency a curve with a maximum at a frequency ω_0 is obtained (*Figure 7.3*), where

$$\omega_0 a = 1 \qquad (7.16)$$

The corresponding maximum of Q^{-1} equal to $\frac{1}{2}(\Delta E/E)$, is known as the 'relaxation strength'; it enables us to

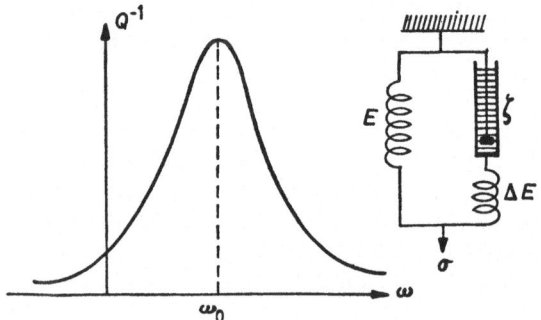

Figure 7.3. Frequency dependence of the damping capacity of the standard linear solid

determine ΔE. Then, with a obtained experimentally with the aid of equation 7.16, ζ may be evaluated (equation 7.14).

Equation 7.16 is particularly useful, for it can often be used to evaluate the activation energy of the process responsible for the damping, and it can thus assist in the identification of the atomic mechanism. This can be seen as follows.

We assume, as an example, that the damping under investigation arises from 'plastic' work due to a limited amount of slip associated with small dislocation displacements about their equilibrium positions. If the dislocation motion is described by equations 5.24 and 4.8, one obtains

$$d\gamma/dt = (1/\eta)\tau \qquad (7.17)$$

with

$$\eta = \eta_0 \exp(Q_0/kT) \qquad (7.18)$$

and

$$1/\eta_0 = 2\rho v_0 b^4 l_j/kT \qquad (7.19)$$

where equation 4.11 was used to yield the specific form of equation 7.19. The temperature dependence of η_0 will in general be small compared with that of η, and if it is neglected, the relaxation time $\eta/\Delta G$ can be written in the form

$$a = a_0 \exp(Q_0/kT) \qquad (7.20)$$

where a_0 is a temperature-independent constant.

If now the temperature of the specimen is changed from T_1 (°K) to, say, a lower one T_2, the frequency of the damping peak will decrease from ω_{01} to ω_{02}; it is readily seen from equation 7.20 that in fact

$$\ln \frac{\omega_{01}}{\omega_{02}} = \frac{Q_0}{k} \left[\frac{1}{T_2} - \frac{1}{T_1} \right] \qquad (7.21)$$

One can therefore determine Q_0 by measuring the frequency of the damping maxima at two temperatures. The method has been used extensively in studies of internal friction phenomena in solids.

7.3 Electrical Analogues

The mathematical treatment of rheological bodies consisting of assemblies of springs and dashpots is analogous to that of electrical L–C–R networks. For example, equation 7.12 is of the same form as that describing the decay of the voltage $V(t)$ across a condenser of capacity C which is being discharged through a resistance R:

$$V(t) = V(0) \,.\, \exp(- C^{-1} t/R)$$

Stress, strain and strain rate correspond to potential, charge and current respectively, while inductance, capacitance and resistance have their counterparts in mass, the inverse of an elastic modulus (a compliance), and the coefficient of viscosity respectively. The electrical analogue of the Voigt–Kelvin body (*Figure 7.1*) consists of a capacitance in *series* with an ohmic resistance, that of the Maxwell liquid (*Figure 7.2*) comprises a capacitance connected in *parallel* with a resistance.

With constant values of the network components the current flow is represented by linear differential or integral equations; analogous equations are obtained for spring and dashpot systems with constant rheological parameters. This constancy of the 'network' elements cannot be assumed with work-hardening materials, except in studies of their anelastic behaviour over narrow ranges of the variables. The structural changes which accompany plastic deformation, such as increases in the dislocation density and hence hardness, in general preclude the use of such differential equations of state even at constant temperature.

In the case of the 'linear', 'viscoelastic' or 'Boltzmann' bodies, consisting of constant m, ζ and E elements, it is possible to deduce useful equations relating to creep, relaxation and deformation under periodic or aperiodic stresses by methods of the operational calculus devised by O. Heaviside towards the end of the 19th century, mainly with the aim of analysing the response of electrical networks to transients. There are numerous textbooks on the subject, and we shall confine ourselves here to a few results of special relevance to the topic under discussion.

Now, the Carson transform $f^*(p)$ of a function $f(t)$ is defined by

$$f^*(p) = p \int_0^\infty e^{-pt} f(t) \, . \, dt; \quad t \geqslant 0 \qquad (7.22)$$

The integral, which represents the Laplace transform of $f(t)$ has been tabulated for a large number of functions.

In shorthand notation equation 7.22 is written

$$f(t) \supset f^*(p) \qquad (7.23)$$

We are particularly interested in the result

$$\frac{d}{dt} \int_0^t f(u) . g(t-u) . du \supset f^*(p) . g^*(p) \qquad (7.24)$$

which we give without proof, and which we shall utilise below.

7.4 Boltzmann Superposition

If a constant stress is applied to the Voigt–Kelvin body (*Figure 7.1*), then from the condition $\sigma(0) = \sigma_E + \sigma_\xi$ one obtains, on using equation 7.3, and 7.4:

$$\epsilon(t) = \sigma(0).[1 - \exp(-Et/\zeta)]/E \qquad (7.25)$$

and for a system of coupled linear bodies one can write, similarly,

$$\epsilon(t) = \sigma(0).f(t) \qquad (7.26)$$

where $f(t)$ is the flow or creep function, and is taken to be zero for negative arguments.

Now suppose that at a time θ after the application of the stress a further increment $\Delta\sigma$ is applied. One then has

$$\epsilon(t) = \sigma(0).f(t) + \Delta\sigma.f(t - \theta)$$

and since the initial extension must necessarily occur in a finite time, it can also be considered for the sake of generality to consist of a series of consecutive stress increments. Hence for any number of changes of stress

$$\epsilon(t) = \sum_{\theta=0}^{t} f(t-\theta).\Delta\sigma(\theta)$$

On replacing the summation by integration, also writing $t - \theta = u$, one obtains the superposition integral

$$\epsilon(t) = \int_0^t f(u) \frac{d\sigma(t-u)}{d(t-u)} du$$

119

As $d(t - u)/dt = 1$, and $f(u)$ is not a function of time, this may be written

$$\epsilon(t) = \frac{d}{dt} \int_0^t f(u) . \sigma(t-u) . du \qquad (7.27)$$

We shall use this result in establishing a relation between creep and relaxation for linear bodies.

We first take the Carson transform of $\epsilon(t)$, which, on comparing equations 7.24 and 7.28, we find to be

$$\epsilon^*(p) = f^*(p).\sigma^*(p) \qquad (7.28)$$

Similarly if, by analogy with equation 7.26, one writes

$$\sigma(t) = \epsilon(0) . r(t) \qquad (7.29)$$

where $r(t)$ is the relaxation function, then

$$\sigma^*(p) = r^*(p) . \epsilon^*(p) \qquad (7.30)$$

It follows from equations 7.28 and 7.30 that

$$f^*(p) = 1/r^*(p) \qquad (7.31)$$

so that if $f(t)$ is known $r^*(p)$ may be found and, consequently, $r(t)$ can be determined. Also, of course, if the relaxation function is known the creep function may be deduced.

For the Maxwell liquid, for example, noting that $\epsilon(0) = \sigma(0)/E$, we have from equations 7.12 and 7.29:

$$r(t) = E \exp(- Et/\zeta) \qquad (7.32)$$

This yields

$$r^*(p) = Ep/[p + (E/\zeta)] \qquad (7.33)$$

and then from equation 7.31

$$f^*(p) = [p + (E/\zeta)]/Ep. \qquad (7.34)$$

Hence one obtains

$$f(t) = \frac{1}{E} + \frac{t}{\zeta} \qquad (7.35)$$

Thus upon application of a constant stress $\sigma(0)$ the strain, as given by equations 7.26 and 7.35, is

$$\epsilon(t) = \sigma(0) \cdot \left(\frac{1}{E} + \frac{t}{\zeta} \right)$$

and is thus seen to consist of the contribution $\sigma(0)/E$ made by the Hookean element, and $\sigma(0).t/\zeta$ contributed by the Newtonian liquid, as was to be expected from elementary considerations.

With a parabolic creep law

$$f(t) = A.t^n, \qquad 0 < n < 1, \qquad A = \text{constant}$$

one finds similarly

$$r(t) = B \cdot t^{-n}, \quad B = (1/A) \cdot \frac{\sin n\pi}{n\pi}$$

8

LIQUIDS

8.1 Viscosity of Simple Liquids and Dilute Colloidal Suspensions

Newtonian behaviour in which the shear rate is proportional to the stress is found in liquids where the shear energy is dissipated mainly by collisions between small molecules. It is observed in water, solutions of inorganic salts, many oils, and in dilute colloidal solutions and suspensions. Suspended matter increases the viscosity. Provided its concentration c, expressed as volume fraction, does not exceed a few per cent the increase is represented by Einstein's formula

$$\eta = \eta_0 (1 + 2 \cdot 5c) \tag{8.1}$$

With high concentrations however η_0 is generally found to become dependent upon the shear rate.

Deviations from Newtonian behaviour may be grouped under two main headings. First we have liquids the rheological behaviour of which can be described fully in terms of the stress and the shear rate, and second, liquids which undergo structural changes in the course of shearing so that the viscosity becomes time-dependent even at constant shear rates. These effects may be reversible; the liquids return to their initial state as soon as the stress is removed or after a certain period at rest.

8.2 Non-Newtonian Liquids

Liquids displaying time-independent non-Newtonian behaviour can be grouped according to whether the shear rate increases with stress faster or slower than for a

Newtonian liquid. The relation between the stress and the shear rate then becomes non-linear. Many equations have been proposed empirically or semi-empirically to describe these phenomena. We shall here utilise only the empirical power-law representation, mainly because of its simplicity. One has

$$\tau = \eta_0 (\mathrm{d}\gamma/\mathrm{d}t)^{1+\delta} \qquad (8.2)$$

where η_0, the 'apparent viscosity' is a constant; it should be noted that its dimension here depend upon the index δ, which makes it rather difficult to ascribe a simple physical significance to η_0 in this representation of the flow.

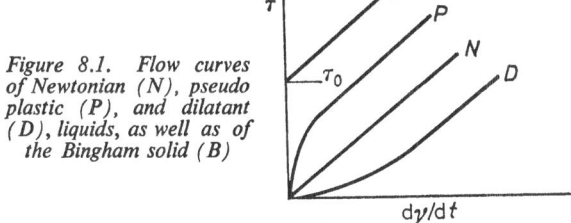

Figure 8.1. Flow curves of Newtonian (N), pseudo plastic (P), and dilatant (D), liquids, as well as of the Bingham solid (B)

If $-1 < \delta < 0$ the liquid is described as 'pseudoplastic' (*Figure 8.1*); if it were truly plastic the flow curve would have to show a definite yield point, e.g. as for the Bingham solid, also indicated in *Figure 8.1*. Liquids for which $\delta > 0$ are generally referred to as 'dilatant'.

Pseudoplastic behaviour is found in solutions of cellulose derivatives and other high polymers, and in suspensions containing elongated particles. These will tend to align themselves at high shear rates so that the long axes or dimensions lie along the stream lines, thus offering a reduced resistance to the shear flow. The most pronounced deviations from Newtonian behaviour will therefore occur at low strain rates.

Dilatancy is often found in concentrated suspensions,

123

such as fine sand in water, or lime-stone dust in bitumen. The packing of the particles becomes less close in rapid shear flow than while the material is at rest and settled, the suspension therefore swells as the shear rate is increased. Energy losses by particle collisions and attrition diminish at the same time and the liquid becomes more nearly Newtonian.

Paints, toothpaste and certain slurries often show Newtonian behaviour only if a certain yield stress τ_0 is exceeded. The rheological flow equation then becomes

$$\tau - \tau_0 = \mathrm{d}\gamma/\mathrm{d}t$$

Such materials are not, strictly speaking, liquids as they maintain their shape at stresses less than τ_0. The existence of the yield stress places them amongst 'plastic' bodies; the simple model behaving as indicated by the line B in *Figure 8.1* was first proposed by Shvedov in 1889, and later studied in detail by Bingham, after whom it is named.

Time-dependent effects occur in thixotropic and rheopectic liquids. In the former a breakdown of the structure occurs on shearing, and the apparent viscosity decreases with increasing shear rate and length of time of shearing. The more linkages are broken the fewer remain intact, available for breakdown, so that at any given shear rate the viscosity attains a lower limiting value approximately exponentially with time. On standing, with the stress removed, the liquid eventually regains its original consistency. 'One-coat' paints, lime pastes and clays, for example, exhibit this behaviour.

In rheopectic materials formation rather than degradation of structure occurs on continued slight agitation; large shearing movements inhibit the effect however. For example gypsum paste containing slightly over 40 per cent by volume of solid in the form of particles with a granulometry of about 1–10 μm, which would solidify in $\frac{1}{2}$–1 hour after shaking if left to stand, can be made to solidify in a few seconds if, after shaking, it is gently rocked in the container.

8.3 Elastic Liquids and the Weissenberg Effect

Viscoelasticity is generally exhibited by concentrated solutions of polymers and other macromolecular substances. A simple model is the Maxwell liquid discussed in Chapter 7. More complex behaviour can be represented by connecting at least one Voigt–Kelvin element in series with a dashpot or Maxwell liquid. The compound body would then display stress relaxation at constant strain and, on removal of the load it would contract somewhat, elastic

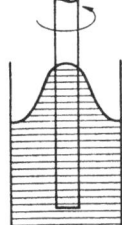

Figure 8.2. Climb of a viscoelastic liquid up a rotating rod, due to the Weissenberg effect

recovery taking place. The occurrence of either or both effects distinguishes a viscoelastic material from a purely viscous, not necessarily Newtonian, one. The elastic contribution to the deformation is responsible for a much discussed phenomenon, known as the Weissenberg effect. One of its manifestations is the climbing of liquid up an inner rod or cylinder of a coaxial system in relative rotation about the axis, as indicated in *Figure 8.2*. The effect does not depend upon the magnitude of the viscosity. It is readily demonstrated with, for example, a solution of a few per cent of polymethylmethacrylate in dimethylphtalate, polyisobutylene in dichlorobenzene, or aluminium laureate in paraffin.

The origin of the effect may be explained as follows. Consider the stress on a small element of thickness dz (*Figure 8.3*) of the rotating liquid, at a distance r from the axis of the system. We take a local set of axes, with σ_{xx}, σ_{yy} and σ_{zz} in the circumferential, radial and axial directions respectively. The stresses will then be as shown in the figure. If the liquid is at rest, i.e. not sheared, the above three stresses are all equal to $- p_0$, the hydrostatic stress at the point of location of the element under consideration.

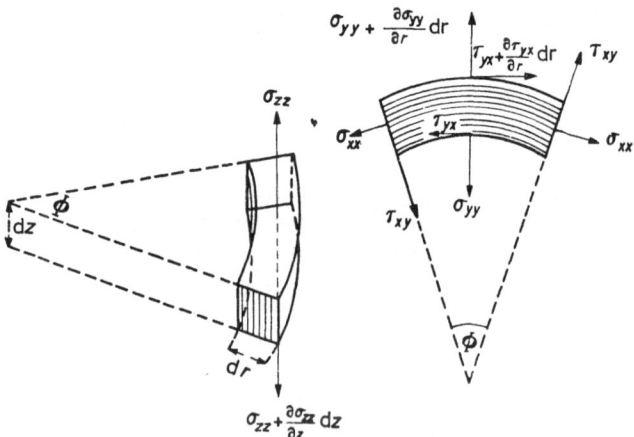

Figure 8.3. Stress distribution in a viscoelastic liquid in laminar shear flow

The sheared liquid may be regarded as bunched elastic fibres aligned, more or less, along the circumferential flow lines, immersed in a viscous, e.g. Newtonian liquid. The appropriate rheological model would then be a Maxwell body. The shear stress τ_{yx} inducing laminar flow will therefore also extend the 'spring element', in this case the fibres, subjecting them to a slight circumferential tension. Assuming the material to be incompressible, the longitudinal extension of the fibres will be accompanied by radial and axial contractions, as if σ_{yy} and σ_{zz} had been changed by equal amounts, both to give a hydrostatic pressure greater than $-p_0$. The circumferential tension, and this effect, may be described formally by writing

$$\sigma_{xx} - \sigma_{yy} > 0 \qquad (8.3)$$

and

$$\sigma_{zz} - \sigma_{yy} = 0 \qquad (8.4)$$

Now, by considering the equilibrium of forces in the direction normal to the shearing planes one finds, from *Figure 8.3*

$$(\sigma_{yy} + \frac{\partial \sigma_{yy}}{\partial r}) (r + \delta r) \, \delta\phi \, \delta z - \sigma_{yy} \, r \, \delta\phi \, \delta z = \sigma_{xx} \, \delta r \, \delta z \sin \phi$$

which yields, as δr and $\delta\phi \to 0$

$$r \frac{\partial \sigma_{yy}}{\partial r} = \sigma_{xx} - \sigma_{yy} \qquad (8.5)$$

In view of equations 8.3 and 8.4 one can infer from equation 8.5 that

$$r \frac{\partial \sigma_{zz}}{\partial r} = \sigma_{xx} - \sigma_{yy} > 0 \qquad (8.6)$$

Since σ_{zz} is negative, i.e. a compressive stress, one can re-write equation 8.6 in the form

$$d | \sigma_{zz} | / dr < 0$$

This inequality implies that in the sheared liquid the absolute value of the hydrostatic pressure decreases as r increases; the pressure on the horizontal surface of the liquid therefore tends to be greatest near the inner cylinder or rod. In the course of rotation a dynamic equilibrium will consequently establish itself, with the higher pressure at the centre balanced by a higher column of liquid near the inner cylinder than at the walls of the container, as indicated in *Figure 8.2.* Equations 8.4 and 8.5 yield the relation

$$- \frac{d\sigma_{zz}}{d \ln r} = \sigma_{yy} - \sigma_{xx}$$

which, together with equation 8.4, has been used in experimental investigations of the Weissenberg effect.

BIBLIOGRAPHY

1. C. A. WERT and R. M. THOMSON, *Physics of Solids*, McGraw-Hill, New York, 1964
2. J. M. ZIMAN, *Principles of the Theory of Solids*, Cambridge University Press, Cambridge, 1964
3. S. TIMOSHENKO, *Theory of Elasticity*, McGraw-Hill, New York, 1934
4. L. R. G. TRELOAR, *The Physics of Rubber Elasticity*, Oxford University Press, London, 1958
5. C. ZENER, *Elasticity and Anelasticity*, University of Chicago Press, Chicago, 1948
6. J. C. JAEGER, *Elasticity, Fracture and Flow*, Methuen, London, 1956
7. A. NADAI, *Theory of the Flow and Fracture of Solids*, McGraw-Hill, New York, 1950
8. D. C. DRUCKER and J. J. GILMAN (Eds.), *Fracture of Solids*, Interscience, New York, 1963
9. F. R. EIRICH (Ed.), *Rheology I*, Academic Press, New York, 1956
10. B. PERSOZ (Ed.), *Introduction a l'Étude de la Rhéologie*, Dunod, Paris, 1960
11. R. HILL, *Plasticity*, Oxford University Press, London, 1950
12. W. OLSZAK, Z. MROZ and P. PERZYNA, *Recent Trends in the Development of the Theory of Plasticity*, Pergamon, Oxford, 1964
13. V. S. POSTNIKOV (Ed.), *Relaxation Effects in Metals and Alloys* (Relaksatsionnie Yavlenia v Metallach i Splavach), Metallurgizdat, Moscow, 1963
14. J. G. TWEEDDALE, *The Mechanical Properties of Metals*, Allen and Unwin, London, 1964
15. A. GELEJI, *Bildsame Formung der Metalle in Rechnung und Versuch*, Akademie Verlag, Berlin, 1960
16. V. A. PAVLOV, *Physical Principles of the Plastic Deformation of Metals* (Fizicheskye Osnovy Plasticheskoi Deformatsii Metallov), Academy of Sciences of the USSR, Moscow, 1962
17. J. J. GILMAN (Ed.), *The Art and Science of Growing Crystals*, Wiley, New York, 1963
18. H. B. HUNTINGTON, *The Elastic Constants of Crystals*, Academic Press, New York, 1958

19. C. R. EVANS, *An Introduction to Crystal Chemistry*, Cambridge University Press, Cambridge, 1964

20. P. H. GEIL, *Polymer Single Crystals*, Interscience, New York, 1963

21. J. FRIEDEL, *Dislocations*, Pergamon, Oxford, 1964

22. M. GORDON, *High Polymers*, Addison Wesley, Reading, Mass., 1964

23. C. KLINGSBERG (Ed.), *The Physics and Chemistry of Ceramics*, Gordon and Breach, New York, 1963

24. YA. FRENKEL, *Kinetic Theory of Liquids*, Oxford University Press, London, 1947

25. W. L. WILKINSON, *Non-Newtonian Fluids, Fluid Mixing and Heat Transfer*, Pergamon, London, 1960

26. A. S. LODGE, *Elastic Liquids*, Academic Press, New York, 1964

27. J. R. VAN WAZER, J. W. LYONS, K. Y. KIM and R. E. COLWELL, *Viscosity and Flow Measurements*, Interscience, New York, 1963

INDEX

Page numbers set in italic type denote most important entries.